Industry and Technology
in Antebellum Tennessee

Industry and Technology in Antebellum Tennessee: The Archaeology of Bluff Furnace

*R. Bruce Council,
Nicholas Honerkamp,
and M. Elizabeth Will*

THE UNIVERSITY OF TENNESSEE PRESS / KNOXVILLE

Copyright © 1992 by The University of Tennessee Press / Knoxville.
All Rights Reserved.
First Edition.

Library of Congress Cataloging in Publication Data

Council, R. Bruce.
    Industry and technology in antebellum Tennessee : the archaeology of Bluff Furnace / R. Bruce Council, Nicholas Honerkamp, and M. Elizabeth Will. —1st ed.
        p.   cm.
    Includes bibliographical references and index.
    ISBN 0-87049-743-X (cloth   )
    ISBN 0-87049-744-8 (pbk.   )
    1. Bluff Furnace (Chattanooga, Tenn.)—History. 2. Blast-furnaces—Tennessee—Chattanooga—History—19th century. 3. Iron industry and trade—Tennessee—Chattanooga—History—19th century. 4. Industrial archaeology—Tennessee—Chattanooga.
I. Honerkamp, Nick. II. Will, M. Elizabeth, 1952-   III. Title.
TN704.U5C63   1992
669'.1413'0976882—dc20   91-32554   CIP

Contents

PREFACE   ix

ACKNOWLEDGMENTS   xiii

CHAPTER 1. ARCHAEOLOGY AND INDUSTRIAL SITES

Industrial Archaeology: A New Subfield   1
The Production and Use of Iron: Beginnings   3
Industrial Archaeology and the Bluff Furnace Site   6

CHAPTER 2. BLAST-FURNACE TECHNOLOGY IN THE NINETEENTH CENTURY

The Raw Materials: Iron Ore, Fuels, and Fluxes   12
The Blast Furnace   20
The Smelting Process   35
Blast-Furnace Operations   37
The Industrial Milieu   41

CHAPTER 3. BLUFF FURNACE AND THE EAST TENNESSEE IRON MANUFACTURING COMPANY

Antebellum Ironworks of Tennessee   45
The East Tennessee Iron Manufacturing Company   50
The Etna Mines on Raccoon Mountain   65
From Charcoal to Coke   67
The Turning Point: November 1860   77
Confederate Iron   85
Reconstruction and Beyond   87
The Archaeological Transformation   89

CHAPTER 4. THE ARCHAEOLOGY OF BLUFF FURNACE

Research Design   95
Methodology   99
The Charging Deck   101
The Steam-Boiler Area   104
The Casting Shed   107
The Furnace Base   112
The Slag Pile   126

CHAPTER 5. BLUFF FURNACE ARTIFACTS

Sampling and Blast-Furnace Archaeology   132
The Artifacts of Industrial Sites   133
The Chemical Analyses   147
Summary   158

CHAPTER 6. BLUFF FURNACE AT CHATTANOOGA: A RETROSPECTIVE

Success or Failure?   161
Chattanooga's Postwar Iron Industry   168
Antebellum Industrialization in the South   175
Bluff Furnace: Past, Present and Future   177

EPILOGUE   181

APPENDICES   183

NOTES   193

BIBLIOGRAPHY   213

INDEX   221

Tables

3.1.	Pig-Iron Production in the United States, 1850	57
5.1.	Bluff Furnace Artifact Groups and Classes	136-138
5.2.	Analysis of Bluff Furnace Pig-Iron Samples	149
5.3.	Analysis of Bluff Furnace Iron-Ore Samples	151
5.4.	Chemical Analysis of Coal Samples	153
5.5.	Chemical Analysis of Bluff Furnace Coke Samples	154
5.6.	Chemical Analysis of Bluff Furnace Slag Samples	156
6.1.	Pig-Iron Production in the United States, 1876	172-173

Figures

2.1.	Sectional View of an Ore Screen	14
2.2.	Section through an Ore-roasting Mound	14
2.3.	A Charcoal Mound	17
2.4.	A Pittsburgh Coke Oven	19
2.5.	Elevation and Section of a Charcoal Blast Furnace	22
2.6.	Section through a Furnace Hearth	23
2.7.	Elevation and Section of a French Coke Furnace	25
2.8.	Section and Elevation of the Great Western Blast Furnace	26
2.9.	Wooden Cylinder Bellows Blast Machine	28
2.10.	Vertical Iron Cylinder Blast Pump	30
2.11.	A Rotary Blast Machine	31
2.12.	A Metallic Water-cooled Tuyere	32
2.13.	Sectional Views of Hot-blast Stoves	34
3.1.	Chattanooga and the Tennessee River Valley in 1860	51
3.2.	Robert Cravens and James A. Whiteside	52-53
3.3.	A Plat of Antebellum Chattanooga	59
3.4.	Ross's Landing at Chattanooga	60
3.5.	Bluff Furnace, 1858	64
3.6.	The Chattanooga Foundry and Machine Shop	68

3.7. An 1860 Stereoscopic View of Bluff Furnace   73
3.8. Union Troops Cross the River into Chattanooga, September 1863   82
3.9. The "Lime Kiln" at the Bluff, 1864   84
3.10. Bluff View in 1905   90
4.1. The Menzi-Muck Climbing Hoe   94
4.2. The Doorway in the North Casting-Shed Wall   96
4.3. Plan of Excavations and Composite Feature Map   98
4.4. Hand Excavation in the Furnace Area   100
4.5. Stone and Brick Construction below the Charging Deck   102
4.6. Anchor Bolts at the Top of the Bluff   103
4.7. Brick Structures in the Steam-Boiler Area   105
4.8. Stratigraphy inside the Casting Shed   108
4.9. West Doorway of the Casting Shed   111
4.10. Two Furnace Bricks   112
4.11. Heavy Structural Cast-Iron Pieces from the Furnace Area   113
4.12. Initial Exposure of the Cupola Base   114
4.13. The Cupola Base after Excavation   115
4.14. Overview of the Cupola Base   116
4.15. The Forehearth Area   117
4.16. Brick Wall and Tuyere Pipes   120
4.17. Composite Map of Furnace-Area Features   122
4.18. Detail of Features South of the Furnace Base   124
4.19. Counterweight Shaft at the Base of the Bluff   125
4.20. Soil Profile North of the Casting Shed   127
5.1. Wrought-Iron Tools   139
5.2. Miscellaneous Fasteners   140
5.3. Fasteners and Fittings   141
5.4. Machine Parts and Fittings   142
5.5. Structural Cast Iron   143
5.6. Two Castings   144
5.7. Pig-Iron Bars   145
5.8. Green Slag   146

Preface

Archaeologists explore a tangible past consisting of altered landscapes, scattered artifacts, and ruined, abandoned structures. In itself, this archaeological record is voiceless; the functional context and meaning of an artifact or building foundation are not always apparent. For the researcher working with the archaeological record of literate cultures, however, the interpretation of the past is framed by written—often subjective—observations recorded by members of that society. In this case, our understanding of remote events and peoples comes from balancing the testimony of documentary history against empirical data collected using scientific methods.

The role of the archaeologist is not that of participant or voyeur, for what we seek to understand has already happened. The events we investigate may have occurred in our lifetime or more than a dozen millennia in the past; the methodology applied is the same. We observe, record, and preserve an account of the past that is as free as possible from the biases imposed by our own beliefs. As a method of inquiry, and distinct from its end products—namely, bodies of knowledge of specific periods or peoples—archaeology encompasses the entire vista of human activity, wherever and whenever it has occurred.

Industrial archaeology is a relatively new scientific discipline. It explores the history of industry and technology through the examination of material remains. The changing focus of the discipline ranges, for example, from the functional characteristics of a specific type of nail to the engineering of the machine that made the nail, from the construction and design of the building that housed the nail-making machine to the geographic setting of the nail factory and its relationship to residential neighborhoods and other industries. Enveloped by industrial archaeology are aspects of the allied disciplines of anthropology, political science, economics, geography, and engineering. Practitioners of this field merge the often sketchy and idealized historical documentation of industrial enterprises with the objective testimony derived from surviving physical evidences of those enterprises. The yield of such studies usually goes beyond a generalized history of industry and technology; combined archaeological and historical data can produce a well-grounded understanding of the development of complex industrial processes, reveal the life-style of the industrial worker and the environment of the workplace, and explore the often curious relationship between the structure of society and its material, economic basis.

The antebellum South is unique in American history in that its economics and politics departed significantly from those of the nation as a whole. The point of divergence was the southern commitment to slavery and to commercial enterprises in which this captive labor force could be efficiently employed. Images of southern agrarianism based on slave labor and aristocratic social structure in the antebellum era are as pervasive as ever in the American mentality. This characterization of the region's past is largely subjective and certainly incomplete. Absent from mental landscapes of the prewar South are industrial facilities such as blast furnaces and rolling mills, coal mines and rail yards, textile mills and lumber yards.

This volume describes an archaeological research project on the remains of an important antebellum blast furnace located in Chattanooga, Tennessee. Formed in 1847, the East Tennessee Iron Manufacturing Company was typical of similar companies of the day. Iron ore was mined in shallow pits from local deposits, and the abundant forests of the surrounding hills were harvested for the production of charcoal. Burned in carefully constructed blast furnaces, the raw materials yielded molten iron that could be either poured into pig-iron ingots for sale in distant markets or cast directly into items for local consumption. The enterprise was expensive and risky. Large capital investments were needed to erect the specialized facilities associated with the furnace, and there were immediate safety hazards as well as long-term health risks for furnace workmen. By nature, ironmasters were conservative, adhering to tried and proven methods virtually unchanged for nearly two centuries. But economic pressures and technological improvements that developed abroad tempted many ironmasters to gamble on new smelting procedures.

The construction in 1854 of Bluff Furnace on the banks of the Tennessee River near the pioneer settlement of Ross's Landing was the first heavy industrial plant in territory wrested from the Cherokee Indians only fifteen years earlier. The furnace took its name from the high limestone cliffs that overlooked a primitive ferry landing at the heart of the settlement later known as Chattanooga. In 1858, a two-year modification of the physical plant of this traditional charcoal-fired blast furnace was begun, resulting in the creation of one of the most advanced iron-smelting facilities in the South. The production of pig iron with coal-derived coke as fuel was successfully tried at Bluff Furnace in the summer of 1860. This first "blast" marked the introduction of modern methods of iron production in the southern Appalachian iron region.

Technological innovations at the furnace did not occur in a vacuum, however. Two northern ironmasters were responsible for the plant conversion, and the success of coke-iron production was not merely a matter of constructing a new facility. In the midst of political turmoil following the election of Abraham Lincoln, the furnace operation collapsed. One of the northern ironmasters fled north in response to local hostility. The Civil War began. Prior to the Federal occupation of the town in 1863, the salvageable equipment of Bluff Furnace was moved to Alabama. Federal troops used the furnace stack as a lime kiln and demolished everything else. By the end of the Civil War, the South's most sophisticated iron-smelting facility was a heap of rubble.

These were the known historical facts about Bluff Furnace, the epochal furnace operation that literally vanished from view until the late 1970s, when an erosional ravine from a storm sewer began to expose a limestone wall foundation, discarded furnace slag, and broken bits of iron machine parts. These physical traces of Chattanooga's first heavy industry were noted by University of Tennessee researchers preparing an environmental impact statement on lands adjoining the riverbank. This chance encounter started in motion a concerted research effort to reclaim an important part of our industrial past.

Armed with what little background information there was, the late Dr. Jeffrey L. Brown excavated test pits at the site and began combing documents for references to the furnace. In 1977 he published a summary of his findings in a small brochure. *Exploratory Excavations at the Bluff Furnace Site* was printed with funds provided by the Chattanooga chapter of the Association for the Preservation of Tennessee Antiquities. Impressed by the site's archaeological and historical potential, the association then hired a local engineering firm to produce a master plan for developing it as a historical park.

Interest in the Bluff Furnace discovery continued to grow. In 1978 the site was nominated for inclusion in the National Register of Historic Places, and it was placed on the register in 1980. The American Foundrymen's Association became actively involved in the project and began raising funds to support continued archaeological explorations. The city of Chattanooga, which owned the property, cooperated by granting permission for further testing and, later, by diverting the storm-sewer outflow that had begun to seriously harm the site's remains. In 1979 the research received a major boost with the formation of Bluff Furnace of Chattanooga, Inc., a nonprofit support organization. Donations of funds and in-kind services were received from individuals and companies, and a substantial grant was awarded to the project by a local charitable foundation.

From the first tantalizing glimpses of the site remains in 1976 until large-scale excavations in 1981, more and more probing questions were asked about the significance of Bluff Furnace in the history of Chattanooga and its role in the industrialization of the South. These questions looked below the surface of the historical accounts and into the broader technological and economic forces that led to the creation of the furnace in its early charcoal form and later compelled its conversion into a coke-fired facility. In answering these questions, we can hope to gain a new perspective on the nature of industry and enterprise in the antebellum South.

Acknowledgments

Archaeological research at the Bluff Furnace site was the result of an intensive effort involving a wide range of individuals and a diverse group of private companies and foundations.

Two archaeological field-schools and one hired professional crew participated in the excavations; we wish to personally recall with fondness, as well as acknowledge the efforts of, the field technicians. From the 1981 field-school and excavation: Alan Ball, Cathy Holt, William Blaine Hunt, Pat Royal, Robert Starbird, Thomas Talbott, Rebecca Terry, Carla Yount, and Sheron Yount; the laboratory personnel were Tommy Talbott, Diana Werner, and Sheron Yount. Excavators in the 1983 field-school were: Nancy Byrne, Laurie DeBoer, Carol Dickert, Bill Ewing, Robert Lambdin, Linda Lancaster, Sandra Moore, Pat Royal, Corban Stone, and Beth Temple.

Organizations and private companies that promoted the excavation of Bluff Furnace, and materially aided us toward that end, were: Bluff Furnace at Chattanooga, Inc.; the city of Chattanooga; the American Foundrymen's Association; the Association for the Preservation of Tennessee Antiquities; the Lyndhurst Foundation; the Campbell Construction Company; the Chattanooga Coke and Chemical Company; Hensley-Schmidt, Inc.; the U.S. Pipe and Foundry Company; Don J. Phen Landscaping; Combustion Engineering; Hunter Museum; the James Wilson Company; the Jones-Haley Construction Company; and Climbing Hoe of American, Ltd.

Numerous individuals aided in the archaeological fieldwork at Bluff Furnace, and still others provided historical information or research services during the writing of this book. We wish to thank the following: Dr. James W. Livingood, Hamilton County historian; Jim Werner, one of the founders of Bluff Furnace of Chattanooga, Inc.; Steve Campbell, formerly of Campbell Construction Company; Kendal Morton, formerly of Chattanooga Coke and Chemical Company; Tim Young, student and occasional laborer; and Sam Rogers, architect. Lance Metz, historian with the Canal Museum–Hugh Moore Historical Parks and Museums, Easton, Pennsylvania, kindly provided us with important historical documents relating to Bluff Furnace, as did Jack Shelley of the Roane County (Tennessee) Historical Commission. We appreciate their contributions. Clara Swan, head of the Local History and Genealogy Department, Chattanooga-Hamilton County Bicentennial Library, and Neal Coulter, Reference Librarian, Lupton Library, University of Tennessee at Chattanooga (UTC), are two persons who materially aided in the progress of our research.

The generosity of Dr. and Mrs. Howard Brown created the Bluff Furnace Gift Fund, from which support for the production of this work was drawn. Dr. Marvin Ernst, UTC associate provost of graduate studies, research, and program evaluation, provided financial assistance in the production of graphics, and Sue Goodwin of the UTC Center for Excellence in Computer Applications graciously allowed use of CECA facilities in the production of the manuscript.

Colleagues John White, Samuel Smith, and several anonymous reviewers read various versions of this volume, and we appreciate their stylistic and substantive criticisms. Alexa Selph, a free-lance copyeditor, made valiant efforts to polish the rough prose of a technical archaeological report and focus the content of the work for the interested general reader. We acknowledge her valuable assistance with this task.

The staff of the University of Tennessee Press remained patient throughout an extended period during which a very rough manuscript was rendered increasingly readable. We want to thank: Carol Wallace Orr, Lee Campbell Sioles, Dariel Mayer, Jennifer Siler, and Stan Ivester.

Finally, we want to recall an ending and a beginning. Through the efforts of the late Dr. Jeffrey Lawrence Brown, first director of the Institute of Archaeology at UTC, the archaeological site of Bluff Furnace was recognized, resulting in its preservation. Jeff's initial excavations at the site were the first long strides toward salvaging an important part of Chattanooga's history and a piece of Tennessee's industrial past. He remained personally involved in directing the future of the archaeological site until death separated him from that task. His scholarly concerns for the site, and the anticipated joys of fieldwork there, remained with us throughout the project. This book is respectfully dedicated to his memory.

Industry and Technology
in Antebellum Tennessee

CHAPTER 1

Archaeology and Industrial Sites

Industrial Archaeology: A New Subfield

Those sites containing the physical remains of former industries and communications have, over the last two decades, become the focus of a new subfield in archaeology known as industrial archaeology. This relatively young addition to the field[1] is closely allied with historical archaeology, itself only about thirty years old. The "historical" dimension associated with both of these branches of archaeology means that their practitioners avail themselves of two types of information: the first, consisting of above-ground data, includes the physical remains at a particular site, along with contemporary corporate or government records (such as censuses or business directories), newspaper articles, photographs, drawings, deeds, diaries, etc.; the second source of data derives from material buried at a site and recovered in the excavation process. Occasionally, historical archaeologists can even include oral information from informants who worked at, or lived near, a particular site. The concentration on "former industries and communications"[2] distinguishes industrial archaeology from historical archaeology in general and likewise requires that its researchers possess a firm working knowledge of the esoteric details of the industrial process under study. This seemingly narrow focus can encompass an astonishing array of sites, from military shipyards to lime kilns, from railroad terminals to umbrella factories. The age of the site investigated depends on the type of industry it represents: industrial iron sites go back several centuries in Britain, for example, whereas "early" nylon factories in that country might date to the 1940s. Not only are some industrial sites still standing, some are still being used!

At first glance, this may seem like a curious and divergent path, one that leads away from what is normally thought of as the traditional area of interest for those who label themselves as archaeologists (i.e., the excavation and study of ancient, exotic civilizations, far removed from ourselves in both time and temperament). A fair question to ask, then, is: Why bother? Do such recent things as nylon factories qualify as archaeological sites? If some industrial sites are still standing, why not just study them instead of painstakingly reconstructing behavior from the bits and pieces in buried sites? And for sites that are no longer extant, why go to the expense and effort of excavation if everything we need to know is written down?

While you may never hear such questions being asked at the yearly meetings of the Society for Industrial Archeology, they nevertheless deserve a response. Industrial archaeology is essentially the archaeology of the workplace, which for most people is the focus of much lifelong effort and attention: roughly one-third of our lives is spent there. The recent age of some industrial sites is immaterial as long as useful information can be gleaned from them. The range of human behavior through time and space is exceedingly wide, and no single period, remote or recent, has a monopoly on all the data necessary for understanding that behavior.

For industrial sites, the sin of omission (missing data) is frequently a more serious problem than the sin of commission (unrepresentative data). For example, what could be more practical, down-to-earth, and ultimately mundane than the upgrading of, say, part of a potash factory? Is this something that would command the attention of the press or historians of the day? If plans had been drawn up for the proposed changes, and if the plans survived, they would in all likelihood bear some resemblance to what was actually done in the factory. But would blueprints have been drawn up at all? Would they record all the modifications that were made, including last-minute revisions? And once the upgrade was complete, would the blueprints have been stored in perpetuity for the edification of future industrial archaeologists and historians of technology? The most reasonable and likely answer to most of these questions is no. That leaves archaeology to fill in the documentary gaps.

There is also an important practical reason to investigate industrial sites: they are disappearing at an alarming rate. Industries themselves have a distressing tendency to destroy their own history through modernization and rebuilding programs, leaving no trace of earlier manifestations. Old facilities get in the way of the new and have to be removed for the sake of economy and efficiency. Entire plants are being cut up and sold for scrap as the U.S. economy continues its shift away from the manufacture of durable goods and into the service sector. The height of industrial production for this country was in the early 1950s. Now, over two-thirds of the American work force is engaged in service or information jobs. This trend is likely to continue, as is the destruction of the physical remains of obsolete or unprofitable industrial facilities. Buried industrial sites, indeed all archaeological sites, will fare no better as the ravages of time, parking lots, and shopping malls continue to take their toll on the archaeological record.

Probably the most important reason to dig into the ground as well as into the documents is that excavation provides two things that are often missing from historic records: context and authenticity. At most sites, serendipity is the rule rather than the exception for archaeological discoveries, as unexpected finds give fuller meaning to above-ground data and broaden and challenge initial interpretations. A long-term perspective is required to make sense of any site. Artifact collections usually relate to continuous events that occurred over an extended time while primary documents commonly contain information about events of relatively brief duration. The context that archaeology provides is by definition a product of time's arrow and requires that we consider how things happened *through* time as well as what happened at any one time.

Industrial and historical archaeologists deal with the tangible, authentic physical remains associated with the past, not just written descriptions or re-creations of it. One of the most exciting aspects of archaeology for all fieldworkers is the physical contact with the reality of the past, literally getting one's hands dirty in the nitty-gritty of history. This facet of archaeology has a powerful appeal, and it presents unique opportunities for interpreting archaeological and historical sites to the interested public, a theme we will return to in the last chapter of this book. By contributing to more accurate and meaningful interpretations—that is, by providing context and authenticity—archaeology ultimately enhances our understanding and our appreciation of the richness of the past.

The Production and Use of Iron: Beginnings

Few would argue the primacy of iron in the shaping of so many aspects of our lives: it provides, literally, the structural foundation of modern urban society. Yet most of us know little about either how iron is produced or how we came to produce it.

Iron and steel are such ubiquitous elements in modern material culture that it is tempting to view them as requirements for *any* successful human endeavor. Metallurgy was a relatively late achievement, however: iron has been around only for about 3,000 or 4,000 years, and large-scale steel production is less than 150 years old. Prior to the introduction of metallurgy, most people did pretty well without metals of any kind, for many millennia.

The preeminence of iron and steel in our present society accounts for the belief that metals have always accompanied civilization. For the Old World, this correlation is fairly accurate, although there the first and most extensive "civilized" use of metal in the Bronze Age was to fashion weapons of war.[3] For the civilizations of the pre-Columbian New World, however, there is little correlation at all. In the northern hemisphere, Native American use of metal was limited to cold-hammering techniques, applied mostly to copper for producing objects of ceremonial, rather than utilitarian, purposes. In South and Central America, some smelting of bronze and other copper alloys occurred, not to mention the artistic manipulation of gold and silver, but it was practiced on a relatively limited scale.[4] The great Mayan, Aztec, and Incan civilizations that flourished for centuries prior to the arrival of European explorers did so with weapons, utensils, and building materials composed primarily of wood, flint, bone, shell, obsidian, etc.—in short, anything *but* metal.

On the other hand, one of the reasons that the vastly outnumbered Spanish conquistadores were able to overthrow entire Indian empires in South and Central America was their use of a military technology based on metal swords, guns, cannons, and armor.[5] Clearly, metallurgy can confer advantages to groups possessing it, even if those advantages are limited to martial endeavors. This seems to have been the case for the earliest known use of iron, by the Near Eastern Hittites in the second millennium B.C. They jealously guarded the secrets of ironworking for some time, and the weapons and tools they fashioned from the new metal helped them extend their political influence over a vast area. For the other complex, highly populated societies of the Near East, Asia, Africa, and Europe, metals were readily adopted once they were invented or introduced through trade or diffusion.[6]

It was not until the Industrial Revolution of the eighteenth and nineteenth centuries, however, that a permanent emphasis on iron and steel began to be felt, as steam-powered machines increasingly came to replace those powered by wind, water, animals, and human muscles. This revolution in energy and work has had profound technical, social, and ideological consequences that forever changed the pace and quality of life for millions of people.[7]

The precise date and location for the European introduction of iron smelting in the New World are unknown, but if we exclude the sixteenth-century smelting experiments in the Canadian Arctic (or the controversial claims for earlier Norse iron production), Virginia becomes our "official" candidate: in 1619 the colony's first blast furnace was thought to be erected at Falling Creek, only to be destroyed in an Indian raid two years later.[8] New England soon became the focus of colonial iron production,[9] and during the eighteenth century Pennsylvania became the preeminent producer. Like most of the other early colonial installations, the Pennsylvania forges were set up by British speculators and investors for the export of iron to the mother country rather than production for domestic consumption. But, by the mid-eighteenth century, the thirty-four furnaces and forges making up the Pennsylvania iron industry had become locally oriented due to the development of large population centers where finished iron products could be marketed.[10] Pennsylvania led the way in iron production, firmly establishing the northeast United States as the locus of heavy industry by the nineteenth century.

Ironically, while southern colonies produced some early iron ventures,[11] they soon lagged behind those in the North. Due largely to its early agrarian emphasis, with the relatively smaller populations and diminished markets for manufactured goods that a diffused agrarian settlement pattern entailed, southern iron companies were at a competitive disadvantage with their northern neighbors. This slowly changed in the nineteenth century, as population centers began to expand and as railroads, critical in both transporting and consuming iron, started to develop throughout the South. Once the local demand for iron products was established, southern (and northern) entrepreneurs took advantage of the natural conditions conducive to iron production. Although not commonly recognized as such, the production of iron was a growth industry in the South until just before the Civil War.

Iron has been around a long time, in this country and others, and it has always assumed an important and inextricable role in the lives of its users. As iron has played, and continues to play, a vital role in our society, the history of its development may shed light on our own historical trajectory into the post-industrial age.

The focus of this book, then, is on iron; more particularly, on the production of iron at an unusual site in antebellum Tennessee. The site contains unique, authentic, tangible remains of Chattanooga's earliest heavy industry, and it represents an important historical example of a fundamental industrial component of modern life. Few sites of this age and type have been thoroughly investigated, and almost none appear in the South. Documentary data have been located—including contemporary descriptions, an 1858 woodcut illustration in *Harper's* magazine, and a photograph of the furnace in operation during its heyday—that can tell us a great deal about the antebellum furnace. Yet, while these seemingly rich sources of information are necessary for understanding the full significance of this site, they are not sufficient. The archaeological record offers considerable insight into the human behavior that occurred at Bluff Furnace through evidence that, combined with the documentary record, leads to conclusions that neither source of data could provide alone.

Bluff Furnace derives its name from its location on a prominent limestone bluff overlooking the Tennessee River in the center of Chattanooga. This site was, over a century ago, a large iron-producing facility on the forefront of iron technology. As the young city of Chattanooga's first and largest heavy industry, Bluff Furnace formed a historical cornerstone for urban development in the region, one that had important ramifications for the industrial emphasis that the city has, until recently, always possessed. No site exists in a vacuum, however. The significance of Bluff Furnace extends beyond local historical particulars to larger political, economic, and technological forces that shaped the South and the rest of the United States during the mid-nineteenth century. In presenting our story, we will examine both the particular site characteristics and the cultural and technological milieu in which these particulars find their fullest meaning.

Industrial Archaeology and the Bluff Furnace Site

How did Bluff Furnace end up as a forgotten part of Chattanooga's buried past after serving as its first major heavy industry? The transformation from a vibrant, powerful blast furnace, complete with shouting workers, a huge, noisy steam engine belching black smoke, and liquid channels of red-hot iron, to a silent archaeological site covered by grass and trees was surprisingly fast. Built in 1854 and shut down just prior to the Civil War, much of the plant was dismantled and its machinery moved elsewhere prior to the Federal occupation of the city in 1863.[12] In the years that followed, eroding soils from the hill above the abandoned site slowly began to accumulate over what was left of the furnace remains. The construction of numerous terraces and houses on the hill deposited additional fill. Other human activities contributing to the process included the trash and debris that were apparently thrown down the hillside, a common disposal practice during the nineteenth and early twentieth centuries. The construction of the adjacent Riverside Drive in the late 1960s applied the coup de grace to the site, as several cubic yards of fill were added to the already substantial overburden.

The old adage "out of sight, out of mind" was never truer than at Bluff Furnace. As its remains slowly disappeared from view, so did its place in history. Finally, except as an occasional footnote in brief discussions of early Chattanooga's beginnings, Bluff Furnace became archaeologically and historically invisible. No physical trace of the site remained to indicate its location at the foot of the Walnut Street Bridge in downtown Chattanooga.

Ironically, the threat to another significant industrial site in Chattanooga led to the discovery and excavation of Bluff Furnace. In the late 1970s, Walnut Street Bridge (built in 1891 and eligible for inclusion in the National Register of Historic Places) was closed, and its demolition and replacement were proposed. Since this project required permits from the federal government, preparation of an environmental impact statement was also required. As part of this project, Dr. Jeffrey L. Brown[13] was hired in 1977 to carry out a ten-day archaeological survey of the areas adjacent to the bridge landings to determine if any significant archaeological remains were likely to be disturbed or destroyed by the demolition and construction activities. Of particular interest to Brown was the south end of the bridge, where a prominent limestone outcropping known as Bluff View appeared. This area aroused his suspicion because he had come across historical documents that mentioned a large bluff near the Tennessee River as the location of an "old furnace." During a surface survey of the property next to the bluff he noticed that an erosional ravine created by runoff from a storm sewer had exposed a massive masonry wall foundation. Also exposed in the profile of the deep gully was a deposit of slag (a by-product of a blast furnace). In addition, Brown discovered numerous iron anchor bolts in the top and sides of the limestone bluff, and his excavation of five test pits produced more evidence of structural footings and architectural debris, confirming that a site was actually present.

At this point, Brown did what any good industrial archaeologist should do: he carried out an extensive review of the literature, searching for information on similar types of sites that had already been excavated and studied. He found disappointingly little comparative material, and over a decade later there is not much to add to it. While numerous historical studies have been completed on various aspects of the early iron industry in the United States, very few have combined archaeological and documentary data, with almost none dealing with sites located in the South.

This is unfortunate. Archaeology is predicated on discovering and interpreting context; individual artifacts, isolated features, and even entire sites have little meaning unless they are compared to, or associated with, other artifacts, features, or sites. In our search of the archaeological and historical literature of the antebellum iron industry, little comparative information was found. One site close in time and space to Bluff Furnace was the Tannehill Furnace complex in northeast Alabama.

Now a state park, Tannehill was built sometime between 1859 and 1861 and was originally a water-powered, cold-blast furnace and foundry worked with slave labor. In 1864 a double furnace powered by steam was erected, only to be destroyed a year later by Federal troops. The site was extensively excavated in 1956, and a short but profusely illustrated report was produced.[14] Although it concerns the first and one of the few systematically studied blast furnaces in the South, the Tannehill report was never published. An excellent synopsis of the site's documented history was published in 1986, but only a short, nontechnical summary of the archaeology was included.[15]

In Tennessee, the Division of Archaeology of the Department of Conservation has published the results of an exhaustive survey of iron-industry sites in the Western Highland Rim.[16] Besides a county-by-county listing of iron-related sites that was generated using documentary and archaeological data, this comprehensive report succinctly summarizes iron-manufacturing technology and the history of industrial development in the highlands region. While not many of the reported sites were fully excavated, this publication does provide comparative data useful in the analysis of some of the Bluff Furnace material.

More typical of the literature on the southern iron industry are occasional articles in history journals, but, not surprisingly, these are based exclusively on documentary, pictorial, or oral data. Les Crocker provides an example of this type of research in "An Early Iron Foundry in Northern Mississippi," which is primarily an economic history of the Holly Springs Ironworks (1839-62).[17] Although Crocker mentions iron artifacts that are "believed" to have been cast at the Holly Springs plant, no systematic below-ground study of the site has been attempted.

As might be expected, industrial development in the Northeast has been studied more extensively than in the South. Besides various sites recorded in the Historic American Engineering Record (a federal program that documents significant industrial facilities and structures), there are a few published archaeological studies of iron-producing sites. An example of the latter is John R. White's important work showing how the analysis of such mundane artifacts as slag and coal samples from the Hopewell Furnace can result in surprisingly valuable information on blast-furnace operating conditions.[18] Such occasional exceptions notwithstanding, it is still fair to say that most of this research has been exclusively document-oriented; *archaeological* studies of blast furnaces are conspicuous by their absence. The few that do exist frequently take the form of contract reports that rarely see the scholarly attention afforded by widely distributed, easily accessible publications. Part of this problem is simply the result of the newness of the field. For instance, Victor R. Rolando's extensive 1980 study of forge and blast-furnace sites in the state of Vermont contains 259 bibliographic notes, not one of which is an archaeological reference.[19]

A notable exception to this general dearth of blast-furnace literature is Bruce E. Seely's article in *IA*, the *Journal of the Society for Industrial Archeology*, on the Adirondack Iron and Steel Company. In his historical and archaeological reconstruction of a mid–nineteenth-century, charcoal-fired, hot-blast furnace site in upstate New York, Seely dispels the myth of antebellum charcoal-iron producers as conservative and old-fashioned, unable or unwilling to change in the face of rapid technological innovation. The subject furnace company successfully adopted hot-blast technology, originally developed for use at anthracite coal furnaces, for its own charcoal-iron production. Seely details this process of "selective modification" and rejects as simplistic the popular view of technological conservatism as the cause of the eventual demise of the American charcoal-iron furnace industry. His case study illustrates the complexity of technological innovation at an isolated ironworks and frames it in the context of the iron industry.[20]

The general absence of archaeology at blast-furnace sites provided Jeff Brown with a compelling reason to carry out excavations at Bluff Furnace. Using University of Tennessee students enrolled in an archaeological fieldschool, he supervised excavations at the site during the summer of 1979. After seven weeks of intense hand labor in the summer heat, the students managed to dig their way through approximately six to eight feet of accumulated fill to expose a small section of what was assumed to be the furnace foundation. They also noted large quantities of slag and iron ore present in the lower strata of the site. But this effort also clearly demonstrated that a number of serious problems had to be overcome if future excavations were going to be feasible, let alone productive. Hand excavation through deep zones of modern fill was neither interesting nor informative. Huge subsurface boulders hindered the digging, as did the extreme slope of the terrain, which sometimes made fieldwork hazardous. Large trees prevented excavation in critical areas. Further archaeology was going to require a concerted effort at a scale commensurate with the unique challenges presented by this site.

The subsequent project, launched in the summer of 1981 and recounted in the following chapters, involved a two-pronged investigation of Bluff Furnace: both historical and archaeological research. To extract information from the archaeological record of Bluff Furnace, and to interpret the documentary data on the site and its larger historical associations, a frame of reference was needed. This was our first task: to review ironmaking technology at the middle of the nineteenth century—an exercise in context.

CHAPTER 2

Blast-Furnace Technology in the Nineteenth Century

Iron in pure metallic form is one of many metals that do not appear in nature; there is no such thing as native iron. A rare exception to this rule is meteoric iron, formed when a ferrous meteorite is smelted as it passes into earth's atmosphere at high speed. To produce ferrous metal on a commercial scale, ores containing ferrous oxides and compounds must be chemically reduced at high temperatures to yield the familiar hard, shiny, and durable substance known as iron. Further alterations and refinements yield steel, the metal of the modern industrial age.

Smelting is a process involving the heat-induced transformation of raw materials in a specialized structure. The general outline of blast-furnace operation is simple. Iron ore, carbon fuels, and minerals called *fluxes* are loaded into the top of the hollow interior of the blast furnace. The burning fuel, driven by a pressurized air blast, generates high, sustained heat that melts the iron out of its ore. Residual compounds and minerals congeal, floating atop the liquid iron drawn by gravity to the base of the furnace. These impurities are drawn off, and the molten iron is tapped out of the base of the furnace and channeled into molds. When chilled, the iron bars or ingots, known as *pig iron*, are ready for further processing into finished forms.

This sketch does little justice to the scale of blast-furnace enterprises and says nothing of the technical and economic problems involved. The interaction of raw materials, the physical plant of the furnace, and the handling of the blast process is complex. In order to objectively evaluate the significance of the artifacts and features uncovered at Bluff Furnace, we needed a technical context for understanding blast-furnace operations. Therefore, before starting the archaeological excavations at the site, we first conducted historical research on blast-furnace technology, thoroughly familiarizing ourselves with mid–nineteenth-century blast-furnace operations. Fortunately, Jeff Brown had amassed a considerable library of technical literature, much of it published in the last century. Many of the books we needed for our study of blast furnaces were in our own library.[1]

Our survey of blast-furnace technology begins with the raw materials from which iron is smelted. Added in measured quantities called *charges*, the combined fuel, ore, and flux placed in the blast furnace is referred to as the *burden*. The accumulated raw materials are called the furnace *stock*. Gathering these raw materials and maintaining the necessary *stockpiles* of these materials at the furnace plant consume most of the labor and expense associated with the operation of the blast furnace.

The Raw Materials: Iron Ore, Fuels, and Fluxes

The chemical element iron, symbolized *Fe*, is one of the most common elements on the planet, but it is always bonded with other elements into ferrous oxide compounds. Since iron does not appear in the earth's crust in native form, as do gold and copper, it has to be chemically *reduced*, or separated, from its complex mineral matrix to produce "pure," workable iron. The impurities that make up the non-metallic content of iron ore are referred to as *gangue*. In the smelting process, this gangue must be chemically separated from the iron and physically removed from the furnace in the form of a by-product called *slag*. Fluxing agents facilitate this chemical separation.

When determining the suitability of iron ore for use in the blast furnace, several characteristics must be considered. These include: (1) iron content, (2) foreign materials present, (3) particle size and density, and (4) moisture content.[2] All contribute to the reducibility of the ore, that is, the relative ease of chemical separation of the iron from the gangue. The principal varieties of iron ore used in the nineteenth century were hematite, limonite, and magnetite.

Hematite, containing the ferric-oxide compound symbolized Fe_2O_3, contains a theoretical maximum of 69.9 percent iron. The reddish color of hematitic iron ore is the basis for its common nineteenth-century name, dyestone. When viewed with the naked eye, a chunk of this ore exhibits small sparkles of silvery iron in a reddish-brown matrix. By the early twentieth century, most of the iron produced in the United States was derived from hematite iron ores.[3] *Limonite* is a collective term encompassing several varieties of brown iron hydroxides.[4] When pure, limonite ores contain 59.8 percent to 62.9 percent iron content. Although lower in iron content than hematite, these ores were considered desirable because of their high reducibility. *Magnetite* has the highest metallic content of the three ore varieties with a theoretical maximum of 72.4 percent. This ore (chemically symbolized Fe_3O_4) is found in only a few restricted localities in North America. The ore has magnetic properties and is black and metallic in appearance. Because of its rarity, it was not a common blast ore despite its high iron content. Magnetite was the most difficult of the ores to smelt due to its high natural density and low reducibility.[5]

In its natural state, iron ore contains a number of mineral compounds that, in the context of iron smelting, are considered impurities. The chief impurities in ore are silica (SiO_2), calcium oxide (CaO), alumina (Al_2O_3), magnesia (MgO), manganese (Mn), phosphorus (P), sulfur (S), and water (H_2O). Other materials end up in the furnace burden, not as a result of chemical bonding but as a result of the mining and handling of the ore. Many of these impurities are removable only by chemical means during the blast. The impurities introduced into the ore during the mining process, or present in cavities in lumps of ore, include various types of clays and shales. Some ores contain calcium oxide or lime equal in quantity to the percentages of silica and alumina in the ore. These ores are considered "self-fluxing," as the natural lime content is sufficient to remove the silica and alumina during the blast so that no separate flux need be added to the burden.

During the nineteenth century, ores were subjected to varying degrees of preparation to remove some impurities and improve the quality of the ore for blast purposes. Ores were washed and screened after mining to remove small particles of ore and dust which, if loaded into the furnace, would impede the flow of gases through the burden (fig. 2.1). Extraneous materials such as lumps of shale could be manually culled from the ore at this point, but there were practical, economic limits to this hand-cleaning operation. Roasting iron ore improved its chemical quality by driving off several impurities and raising the level of oxidation in the ore. The iron ore was burned in heaps, in long rows, or in specially built ovens (fig. 2.2). The process helped to remove such undesirable elements as phosphorus, chlorine, arsenic, and sulfur.

Ore was size-graded prior to use, and large ore chunks were broken to the desired diameter either by hand or by mechanical ore stamps. Ore used in nineteenth-century furnaces was broken into lumps two or three inches in diameter.[6] Small particles of ore and ore dust could not be used in furnaces until development of a process called *beneficiation* in the late nineteenth

Fig. 2.1. Sectional view of an ore screen. Iron ore was sifted on screens such as this to remove small ore fragments and dust that would clog the interior of the blast furnace. From Overman, *The Manufacture of Iron*, 47.

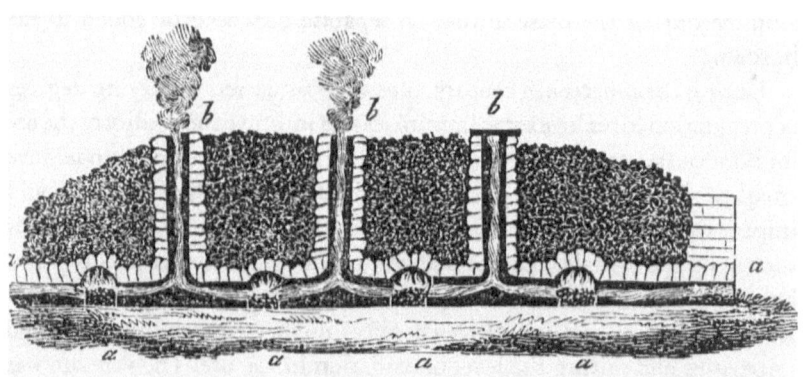

Fig. 2.2. Section through an ore-roasting mound. To dry the iron ore and drive off some of its impurities, it was baked in specially constructed mounds. This cutaway drawing shows the stone flues and vents drafting the heating fires beneath the mound. From Overman, *The Manufacture of Iron*, 44.

century. This process consolidated crushed ore and dust into briquettes of ideal size for blast furnaces.

The element that combusts in the blast furnace is predominantly carbon. This chemical element combines readily with oxygen and, in the process, releases heat. Hydrogen is the other element that burns in the blast furnace. The effectiveness of a blast fuel is mainly determined by its carbon content and the nature of its other chemical constituents. Secondarily, the physical characteristics of a fuel are also important. Air flow through the furnace burden is essential to combustion of the fuel and to the other chemical processes taking place; a charge of good blast fuel should settle in the stack, leaving spaces or interstices through which gases may readily circulate. The fuel must not be easily crushed by the weight of the burden, since this crushing would fill the interstices in the burden with fine debris, blocking gas circulation.

A variety of organic substances contain high percentages of carbon and hydrogen, along with other elements and compounds. Plants in particular have a high carbon content. Thus vegetation, in the form of raw or charred wood or in the form of fossil fuel, was (and is) the dominant blast fuel. Although wood and peat were sometimes used in blast furnaces of the nineteenth century, the principal fuels used were charcoal; *anthracite* or hard coal; *bituminous*, or soft, coal; and coke made from bituminous coal.

All vegetation, even in fossilized form, contains water and dissolved gases. The water and gases are volatiles, that is, they will evaporate from the wood or coal at certain temperatures. To render organic matter into a blast-furnace fuel, these volatiles had to be removed by carefully managed burning, a controlled partial combustion. Thus wood was changed to *charcoal*, and bituminous coal into *coke*. The processes used to accomplish these changes are referred to as *charcoaling* and *coking*. The liberation of volatiles from organic fuels is a natural as well as artificial process. Large deposits of organic material buried in earth sediments have, through millennia, been "cooked," or metamorphosed, by intense geological pressure in a manner chemically analogous to the process of charcoaling or coking. These buried organic deposits have produced carbon-rich deposits of coal. Soft, or bituminous, coal contains tar-like impurities and some volatiles that can be removed by coking. Coal that has been geologically "cooked" the longest is anthracite coal, a hard, dense, and relatively pure coal.

For over half the nineteenth century, charcoal was virtually the only fuel used to smelt iron on a large commercial scale in the United States. In comparison to coals, it is relatively pure in its chemistry. Elements harmful to the production of good iron, such as sulfur and phosphorus, are not present in significant quantities in wood. Freshly cut wood is mostly water, however, and this and other volatiles in the wood must be removed by charcoaling. In the process, the bulk of the wood is transformed into a firm, compact fuel with excellent chemical properties.

Many hard and soft woods were suitable for use in blast furnaces after charcoaling, although hardwoods with high specific densities, such as oak, sugar maple, and beech, were preferred. Woodcutters felled the trees and cut the trunks into lengths of 4 feet. The wood was then stacked in cords for accounting purposes. A cord of wood was 128 cubic feet, or a pile of timber 4 feet wide, 4 feet high, and 8 feet long. An average acre of virgin forest contained about thirty cords of wood.

Colliers, the workers responsible for charring the wood, were specialized technicians and often some of the best paid of the furnace work force. Several methods of charcoaling were available, including charring in heaps or mounds, or in specially built masonry kilns, or *retorts*. Charring in mounds was the most common used in the early nineteenth century (fig. 2.3). A typical mound measured thirty feet in diameter, was ten feet high, and contained perhaps twenty or twenty-two cords of wood. After an initial open fire to start the process, the mound was sealed with earth. Trapped heat expelled gases from the wood. The volatiles consumed themselves, leaving the carbon matrix of the wood intact.

On a larger scale, colliers might also use closed, masonry ovens or kilns to char wood. A typical brick charcoal oven from the antebellum period measured twelve feet high and twelve feet deep by fifty feet long, and could hold fifty to sixty cords of wood. Charcoal ovens were relatively more secure against the vicissitudes of the weather. The manner of charring followed the same basic procedures as those employed in the mounds. Charcoaling would reduce one hundred pounds of wood to approximately twenty pounds of charcoal.[7]

A typical nineteenth-century blast furnace producing two tons of iron a day completely consumed about 250 acres of mature timberland a year. For a furnace to be self-sustaining, several thousand acres of timberland had to be cut in rotation. By far the largest single occupation associated with the operation of a charcoal blast furnace was the cutting and charring of timber. Maintaining a reserve of timberland to provide the needed charcoal was a large investment. As a furnace consumed the woodlands around it, the distances traveled to obtain wood increased, and with it, the cost in time

Fig. 2.3. A charcoal mound. A lengthy and expensive operation, charcoaling involved the controlled combustion of carefully constructed mounds of cordwood sealed by dirt and leaves. By restricting the amount of air to the open fires in the interior of the mound, the moisture was driven out of the wood, leaving the dark carbonized residual, charcoal. From Overman, *The Manufacture of Iron,* 106.

and money of transporting the charcoal. Great care had to be taken in hauling charcoal to the blast furnace, as it was fragile and easily reignited in open air. Teamsters who took on loads of charcoal with hot embers remaining in them delivered burning carts to the furnace.

In the early nineteenth century, ironmasters in the smelting regions of Pennsylvania and New York began to feel the crunch of a shortage of cheap charcoal. The rich iron-ore deposits of the region had attracted the nation's highest concentration of blast furnaces, and the dwindling forests meant competition for a diminishing fuel resource. As charcoal production costs went up, the furnacemen turned their attention to an abundant local fuel—coal. Many furnaces experimented with combinations of blast fuels, usually mixing the ever-more-expensive charcoal with easy-burning soft coal.

Both varieties of coal were present in the Northeast: soft, bituminous coal, and the harder, but chemically purer, anthracite coal. Both were difficult to use in early nineteenth-century furnaces. Bituminous coal contained tar-like substances and impurities that made its use chemically undesirable, although it was relatively easy to combust. Anthracite coal had a better chemical profile, but its higher density made it difficult to burn using the conventional low-pressure cold blast.

The application of the hot blast facilitated the combustion of anthracite coal. By preheating the blast, the hard, dense anthracite coal could be burned easily. The Neilson hot-blast stove, patented in Scotland in 1828, was the first in a series of blast heaters developed in the nineteenth century. The Neilson stove was not in commercial use in Great Britain until 1837, by which date an experimental American hot-blast furnace had already completed a trial run smelting solely with anthracite.[8] Anthracite pig-iron production in the United States commenced on a commercial scale about 1840, although experiments with the fuel, sometimes mixed with charcoal, had begun as early as 1815. Within the geographical limits of the anthracite coal field, the iron-smelting industry of the Northeast shifted rapidly away from the use of charcoal and adopted the readily available, abundant mineral fuel.

Raw coal in its bituminous form was not widely used due to its low porosity and tar-like impurities, although in the late nineteenth century a number of furnaces in the southern Ohio region were burning it. Bituminous coal could be made more suitable as a fuel by altering its chemical makeup, and in its altered form it became a popular furnace fuel. The type of distillation that transformed soft coal into a desirable fuel was known as *coking*.

In a manner analogous to charcoaling, coking removed volatiles from coal, rendering a fuel with not only improved chemical characteristics but an improved physical structure. As volatiles evaporated out of the coal, a hard, bubbly texture was created. The bubbled texture of the fuel served to promote rapid combustion because of a high surface-area-to-volume ratio. When cooled, the hard, porous coke was lighter than coal but also physically sturdy, enabling it to support the weight of the furnace burden without crushing. Good grades of blast-furnace coke contained 85 to 90 percent fixed carbon.[9] When coked in ovens, coal was reduced by 25 to 50 percent of its original volume (fig. 2.4).

By 1854, pig iron produced with anthracite coal exceeded the gross tonnage produced by the traditional fuel, charcoal. In 1869, pig-iron tonnage produced with either raw bituminous coal or coke exceeded the charcoal production. Finally, bituminous coal, predominantly in the form of coke, surpassed anthracite pig-iron production in 1875.[10] Motivating this shift in blast-furnace fuels was simple economy. First, although charcoal was a premium blast-furnace fuel, its physical structure did not permit it to carry the weight of a large burden; consequently, it could only be used in small furnaces. Charcoal was softer than coke and could not be transported great distances or handled by bulk-process machinery. With

Fig. 2.4. A Pittsburgh coke oven. Coal was distilled into coke in a manner similar to charcoaling. While coking could be done in heaps outdoors, a better product was produced inside a masonry oven. This nineteenth-century example illustrates one such oven, built of brick and stone masonry and set in a hillside. The narrow apertures at the top of the oven opened into large circular cavities lined with firebrick. From Overman, *The Manufacture of Iron*, 123.

respect to mass-production methods, then, charcoal had serious limitations. Using coke fuel, furnaces could be built on a much larger scale; the sturdy coke held up well in large burdens.

In the southern iron-ore fields at the middle of the nineteenth century, wood was still plentiful and cheap. The bituminous coals of the South tended to contain more phosphorus and sulfur than their northern counterparts, making their use less desirable than charcoal. Since the use of coke conferred few advantages to southern ironmakers, charcoal pig iron continued to be made in the South long after the North had turned to larger furnaces fueled by mineral coals.[11] The market for southern charcoal pig iron, although small, remained intact until well into the twentieth century. The southern preference for charcoal was principally a rational economic decision, but one with long-term consequences for the iron industry of the region.[12]

Although some ores and fuels contained enough lime to be self-fluxing, most mineral fuels required that a separate *flux* be added to the burden. The fluxes were added to the fuels and ores to perform specific functions, namely, to chemically facilitate the reduction of iron from its oxide compounds in the ore and to collect impurities released from the ore and fuel. The flux became a kind of sponge, acting to congeal impurities such as sulfur, silica, and alumina. The slag that formed could then be isolated from the molten metal and removed.

The principal blast fluxes in use in the nineteenth century were limestone, dolomite or magnesium limestone, and silex. Besides being abundant, limestone and dolomite were considered the optimum fluxing agents. The slags they produced were slightly more fusible than iron itself, thus readily effecting the separation of the iron from the other constituents of the ore.

The Blast Furnace

Iron ore could be reduced in either of two ways in the early nineteenth century, each yielding a specific type of iron. At *bloomary forges,* ore was smelted in relatively small quantities using charcoal as fuel. The molten iron was removed from the hearth of the small forge in a pasty mass called a *bloom* and hammered into malleable *wrought iron*. At *blast furnaces*, ore was smelted on a much larger scale, and the metal tapped from the hearth was pure, brittle *cast iron*. By the end of the century, only blast furnaces were in the business of smelting iron ore; the tiny bloomary forge had long been abandoned.

The crucible in which the smelting of iron ore took place was the furnace proper, or *stack*. The typical early nineteenth-century stack was a giant chimney consisting of an outer structure of heavy stone blocks and an inner chamber lined with delicate refractory stone or clay firebrick. These liners were heat reflecting and heat resistant. The chamber or hollow interior of the stack varied as to basic dimensions and shape, as did the stack structure surrounding the chamber.

The stack was a vertical structure with two main centers of activity. Fuel, flux, and ore—the furnace burden, or charge—were loaded into the top of the stack, also known as the *throat* or *tunnel head*. Near the base, the air blast was delivered, either by steam or water power. At the very bottom of the stack the molten iron and slag were removed. The burden was loaded into the top of the furnace from a platform known as the *charging deck*. It was common practice in the first half of the century to situate the furnace at the base of a hill or bluff where the difference in elevation could be used to advantage. The charging deck ran out from the top of the bluff or hill to the furnace top, creating a pathway and working surface to the tunnel head. The base of the furnace often sat in a hollow carved out of the side of the hill or bluff. The walls and roof of the *casting shed* abutted and surrounded the base of the furnace, creating a sheltered environment in which the molten metal could be cast safe from the elements.

The blast furnace had to be constructed on firm, well-drained ground that could support several tons of masonry and metal. Roofs covered the charging deck at the upper level of the furnace and shielded the stockpiles of materials being loaded into the furnace, but in early furnace practice the actual opening at the top of the furnace stack was often not covered.

The typical furnace at mid-nineteenth century consisted of a massive stone or brick structure in the form of a square truncated pyramid perhaps thirty-five feet high (fig. 2.5). The stone exterior surrounded the combustion chamber, which was lined with refractory material, usually specially made *firebrick* in the upper portion of the stack and cut sandstone blocks near the bottom. Between the lining and exterior structure the furnace might be filled with laid stone, brick, or perhaps stone and brick rubble for thermal insulation. Incorporated into the masonry during construction were wrought-iron restraining bars not unlike the steel rods used in modern reinforced concrete. These restraining bars or *tie rods* served to reinforce and bind the masonry structure together. The shape of the furnace chamber varied considerably, depending on various factors such as furnace height, type of fuel, and ore. But always the furnace chamber was circular in plan, narrow at the top, and broad near the base, the widest point of the chamber being at the top of the *boshes*.[13]

In the interior of the furnace, the *bosh* or *boshes*, was a kind of shelf near the base of the stack created by a sharp narrowing of the chamber above the hearth, the narrowest and lowest portion of the chamber. This shelf had an important function; the sloping ledge served to carry the weight of the burden and acted as a funnel to condense the melting charge into the hearth.

The *hearth*, sometimes called the *crucible*, was the lowest point in the stack, and where the molten iron collected and puddled (fig. 2.6). It was just above the floor of the hearth that the air blast was admitted into the chamber, and consequently this was the hottest point in the furnace. The structure surrounding the hearth region of the furnace was known as the *inwall* or *inwalls*. The hearth was lined with heavy refractory masonry, usually individually cut sandstone blocks. On one side of the lower hearth the lining opened out into the *forebay*. Along the bottom of this opening was the *damstone*, a structure that held back the molten iron in the forebay. Molten iron was tapped out through a hole in the damstone plugged with removable fire-clay balls. Above the forebay opening was the *timpstone*. The

Fig. 2.5. Elevation and section of a charcoal blast furnace. This front view and side cutaway view of a typical nineteenth-century masonry furnace shows the basic exterior and interior shape of the blast furnace. The cutaway view shows the charging deck, and the loading gate at the furnace top. The tuyere and runout arch are shown in the lower portion of the view, as are the shape of the stack chamber and hearth. In the cutaway below, the air-supply pipe, or bustle pipe, is shown buried beneath the base of the stack. From Overman, *The Manufacture of Iron*, 155, 152.

Fig. 2.6. Section through a furnace hearth. The complex masonry of the blast-furnace hearth involved the careful fitting of the: (a) bottom stone, (b) damstone, (c) sidestones, (d) tuyere stones, (e) topstone, (f) tuyere holes, (g) timpstone, and (l) uprights associated with the timpstone. From Overman, *The Manufacture of Iron*, 159.

faces of the forebay, damstone, and timpstone were sometimes revetted with iron plates.

It was to the lowest portion of the stack, at the hearth, that ironworkers needed continuous access to monitor the smelting process, to adjust the nozzles controlling the blast, to draw off slag, and to tap out molten iron. Hence, at the base of the stack, and usually on all four sides of the structure, the massive masonry stack gave way to openings with arched roofs. These openings to the base of the stack were called the *tuyere arches*, so named because through these archways the blast pipes were admitted into the hearth region of the furnace through openings called *tuyeres*.[14] The arch at the front of the furnace, where the iron was actually tapped out, was called the *run-out arch*.

As the nineteenth century progressed, furnaces became larger in size and underwent some stylistic structural changes reflecting the adoption of new materials, construction techniques, and economies of scale. About the middle of the century, a new style of furnace structure was introduced into the United States. The term *cupola blast furnace* is somewhat misleading and considered a technical misnomer by some industrial historians but is nonetheless descriptive of a specific blast-furnace stack design (figs. 2.7 and 2.8). The term *cupola* is used specifically in reference to small iron-shelled furnaces used in foundries to remelt pig iron for casting. The small foundry furnaces were slightly built in comparison to the typical masonry or large cupola-type blast furnace; they were round in plan and used charcoal, coal, or coke as fuel.

The large cupola-style blast furnace was a European innovation that spread to the United States in the 1850s. The refractory lining of the furnace chamber was surrounded by a circular exterior shell of thick cast-iron plates bound together by heavy vertical and/or concentric wrought-iron bands. The tuyere and hearth section of the lower stack was open on all sides, the upper stack being supported in this region by cast-iron pillars topped by a heavy iron plate or *mantle*. This arrangement gave better access to the tuyere and hearth area and made regulation of the blast equipment easier. On the whole, the cupola-style furnace was a more compact structure than the bulky masonry pyramid of earlier designs.

The cupola-style blast furnace did not meet with rapid acceptance after its introduction to the United States, although in Great Britain the use of this type of stack was nearly universal by the middle of the nineteenth century.[15] Some contemporary critics were suspicious of cupolas, complaining that the round stacks "work irregularly[,] ... consume a greater amount of fuel than square stacks[,] ... [and] they nearly always break the strongest binders" due to increased expansion and contraction of the metal stack.[16] Nonetheless, the cupola-type blast-furnace stack was the pattern widely adopted at new furnaces built in the American West beginning in the 1860s,[17] and in the Northeast this style steadily replaced the older massive masonry stacks.

The cupola-style or iron-shell furnace eventually replaced all the archaic square masonry stacks. At the root of this change were economics and technology. Structural iron and steel conferred the advantage of strength to the furnace stack, permitting larger furnaces to be constructed. The sturdy physical characteristics of coke worked hand-in-hand with bigger furnace stacks. Coke-fired iron-shelled furnaces rapidly increased in size, taking advantage of the economy of scale. The typical furnace stack of a coke-fired smelter in the late 1800s was often ninety feet high, three times the size of its charcoal-fired predecessors.

To achieve the high temperatures needed to smelt ore, the fuel combustion in the blast furnace was *pushed*, that is, speeded up by the forcing of air

Blast-Furnace Technology in the Nineteenth Century / 25

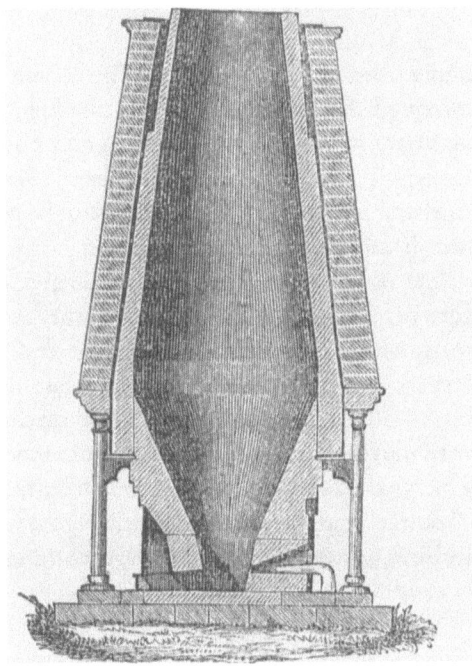

Fig. 2.7. Elevation and section of a French coke furnace. Unlike the massive truncated masonry pyramid of traditional furnaces, the cupola-style stack featured a relatively thin insulating shell. The lower portion of this stack, from the mantle plate down, closely resembles photographs of the lower portion of Bluff Furnace. From Overman, *The Manufacture of Iron*, 177-178.

Fig. 2.8. Section and elevation of the Great Western blast furnace. This example of the cupola-style blast furnace combines the traditional charcoal hearth and tuyere arches with the round upper stack of the cupola. From Overman, *The Manufacture of Iron*, 163.

under pressure into the hearth. The enhancement of the supply of oxygen increased the combustion rate of the fuel, releasing the quantity of heat necessary to initiate the chemical and physical changes in the ore. The forced-air current, known simply as the *blast*, was thus at the heart of the smelting process. The blast had first to be generated, then delivered to the hearth, and be capable of regulation.

The *blast machine* was the mechanical device that created the forced-air current. During the nineteenth century, blast machines underwent dramatic changes in working principle and construction. Other than the earliest and most archaic form of blast machine—the *trompe*, or water blast—all nineteenth-century blast machines were mechanical devices with moving parts that had to be set in motion by some form of externally generated power. At mid-nineteenth century, there were only two choices of power: water or steam. Whether generated by a rotating waterwheel, turbine, or steam engine, the blast machine received its power in the form of torque on a rotating shaft.

Water-power installations were often expensive to build, involving the construction of dams, flumes or raceways, and the creation of water reservoirs, not to mention the cost of the waterwheels or turbines. Once constructed, however, water power was relatively cheap to operate and abundant in most of the iron-producing regions of the country; the technology had been in use for centuries prior to the 1800s. The mechanical power developed by the waterwheel, in the age before electricity, could not be transmitted any appreciable distance. Therefore, the blast machine, and the blast furnace, had to be relatively close to the hydraulic power system. This locational restriction was sometimes at odds with the location of iron ores. River courses and streams were also subject to seasonal variations in flow, and catastrophic floods were altogether too common in the days before watershed management.

Steam power was more expensive to operate but, compared to water power, was more dependable. The steam engine required the constant attention of a new breed of worker, the *mechanic*, and could only be built at specialized industrial plants: the foundry and machine shop. The cost of these expensive machines was counterbalanced by their ability to be situated virtually anywhere that fuel (wood or coal) was available or could be economically shipped, and they were relatively compact as power plants went. This locational freedom became an increasingly strong advantage during the century.

Primitive *blast bellows*, occasionally of enormous size, were largely gone by the nineteenth century. The triangular bellows, opened and closed by a rotating cam shaft, delivered a pulsing blast that was considered unsatisfactory. A more advantageous arrangement was the *cylinder bellows*. This consisted of a cylinder-and piston-arrangement functioning as a single-acting air pump (fig. 2.9). The air cylinders, constructed of wood, were usually arranged vertically in pairs. Pistons were driven vertically through the cylinder, forcing air into a receiving cylinder. Valves on the piston heads and receiving cylinder regulated ingress and egress of the air flow. The air driven from the cylinders came into the receiving cylinder in pulses. A floating, weighted piston in the receiving cylinder was buoyed by these air blasts and acted to regulate the air flow out of the receiver, thus producing a more even blast pressure.

The-cylinder-and piston arrangement, which could be driven by steam or water, later used a double-acting pumping cycle; double valves on the piston and cylinder permitted air to be pumped on both strokes of the piston. Working on a crankshaft, multiple air pumps could fill the receiving cylinder, further evening the air blast (fig. 2.10). Some furnaces had horizontal banks of double-acting pumps driven by water power[18]

Fig. 2.9. Wooden cylinder bellows blast machine. This cross-section of a typical nineteenth-century blast machine illustrates the simple but effective technology used to create a pressurized air blast. Two air cylinders (a) are connected with the receiver cylinder (b) at the top by simple flap valves. The blast pipe (c) runs to the furnace, or, if a hot blast is employed, to the hot-blast stove. Piston heads (d) are driven through a vertical stroke by the pivoted beam (f) working in cam fashion from a tie rod attached to the crankshaft wheel (e). Flap valves on top of the pistons close when the piston is driven vertically, driving air through the open valve in the receiver. On the down stroke, the receiver valve closes and the piston head valve opens. A piston head in the receiver (g) is buoyed by each blast, the air pressure in the cylinder being regulated by a counterweight (h) on a vertical shaft. From Overman, *The Manufacture of Iron*, 395.

while others used steam engines driving vertical pumps. All-metal pumps replaced less durable wooden ones, and because of uneven wear on the air pistons and cylinders, vertical pumps were preferred over horizontal ones.

By the middle of the century, *rotary blast fans* had been developed. The rotating fan mechanism (fig. 2.11) was relatively simple, and its continuous air blast did not require clumsy regulating mechanisms like the receiver cylinder. From simple paddle-bladed fans, which could not deliver an impressive blast pressure, the rotary blast mechanisms evolved during the late nineteenth century into more sophisticated and efficient high-speed air turbines capable of producing high blast pressures. By the late nineteenth century, the largest blast-furnace operations were equipped with multiple steam-powered blast engines generating several thousand horsepower each. Blast pressures increased steadily through the nineteenth century. By way of comparison, the trompe evidently delivered quite a bit less less than one pound per square inch (psi) at the hearth, and the cylinder pump at mid-century delivered perhaps three to five pounds.[19] By the turn of the century, blast pressures of fifteen to thirty psi were not uncommon.

The blast generated by the blast machine was conveyed by blast pipe to the furnace or, if a hot blast was employed, to the hot-blast oven. The shorter the distance the blast had to travel, the less pressure was lost to friction in the blast pipe. Consequently, the blast machine was typically situated close to the furnace. Sharp bends in the pipe were avoided to eliminate obstructions to air flow. In hot-blast furnaces, careful attention to pipefitting was necessary because the hot-blast air caused expansion in the metal blast pipe.

At the base of the furnace, where the blast was delivered into the hearth, the air was conducted to the tuyeres, the openings in the furnace lining through which the blast was delivered into the burden. The number of tuyeres might vary from one to as many as nine, depending on furnace size. In later furnaces, tuyeres were arranged in two banks, one above the other. The objective of situating the tuyere or tuyeres was to distribute the blast evenly into the furnace burden, making certain there were no hot or cold spots in the hearth. Hot-blast furnaces could use relatively lower blast pressures than cold-blast furnaces, given the combustion enhancement of preheating. However, as a consequence of lower pressures, they required more tuyeres to distribute the blast evenly.

The metal blast pipe usually had a large diameter. In a furnace with multiple tuyeres, the blast pipe often circled the furnace either above the level of the hearth or below ground. This *bustle pipe* arrangement evenly distributed the air to each tuyere. Smaller pipes fed off the main pipe to

Fig. 2.10. Vertical iron cylinder blast pump. More durable than the wooden blast tubs were metal blast cylinders. In this example, a pair (only one is visible) of cylinders is used, the pistons being driven by the rocker arms at top. These pumps were double-acting, that is, valves on either end of the cylinder permitted the piston to pump air on the up and down strokes. From Overman, *The Manufacture of Iron*, 398.

deliver air to individual tuyeres. At the end of the feeder pipe was a *blast nozzle*, the terminus of the blast conduit, which took the form of a funnel. The blast nozzle was inserted into the tuyere and sealed in position with fire clay. The feeder pipe and nozzle often curved at a right angle just before entering the tuyere; a plug pipe-fitting at this point permitted access to the interior of the blast nozzle. Since the tuyere might become clogged during the blast, a poker rod could be inserted into the tuyere through this elbow pipe-fitting and the obstruction cleared.

The tuyeres of cold-blast furnaces cooled themselves with the continuous air current of the blast. The preheating of the blast, however, dictated that the tuyeres had to be provided with a cooling mechanism or they would erode and fuse when exposed to high hearth temperatures. The metal water-jacketed tuyere was developed for this purpose. Water contacting very hot metal or masonry could produce scalding steam, and the sudden chilling

Fig. 2.11. A rotary blast machine. This paddle-blade fan mechanism did not produce an impressive blast pressure, but it had the virtue of being a continuous one, unlike the pulsating blast produced by cylinder-type blast machines. This type of blast machine eliminated the need for pressure-regulating mechanisms such as the receiver cylinder. Later examples of this early type of rotary blast machine were designed to produce substantial blast pressures. From Overman, *The Manufacture of Iron*, 408.

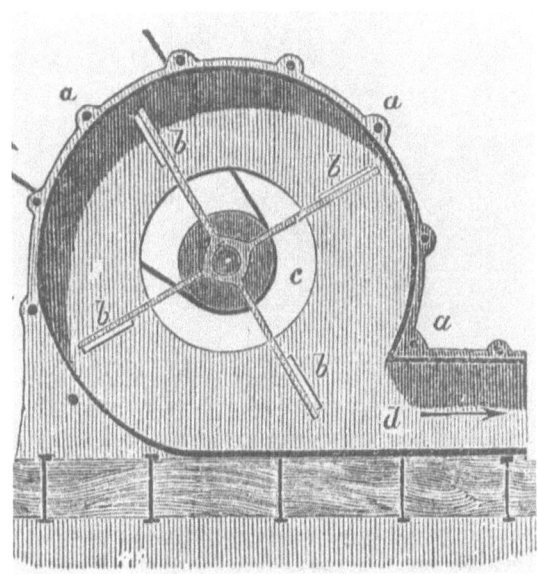

could result in fracturing or spalling. Large volumes of circulating cooling water in close proximity to the super-heated hearth were a constant source of concern to furnacemen.

Several variants of the water tuyere were developed. One popular style (fig. 2.12) featured a metal pipe coil cast into the body of the tuyere. Cooling water was pumped to the nose of the tuyere (the ingress line) where the heat was greatest. As the water spiraled out of the tuyere (the egress line), heat was removed. In another variation, the tuyere shell was hollowed out into a water reservoir, and cooling water was continuously circulated through the shell.

The single most important development in nineteenth-century blast-furnace technology related to the generation and conservation of heat in the smelting process. The principal innovation in heat technology was the hot blast. Stated simply, the hot blast involved preheating the air blast being piped into the furnace through the tuyeres. The preheating of the air enhanced the combustion of the fuel, resulting in greater efficiency and economy in the combustion; the fuel burned faster and hotter. An economy of fuel was obtained, meaning less fuel was needed per ton of ore smelted.

The second aspect of this hot-blast revolution was just as significant; hot waste gas being vented at the furnace top was recycled to heat the blast. The waste gases vented at the furnace top also came to be used to generate steam to run the blast machine. In essence, the blast fed itself in a thermal closed-

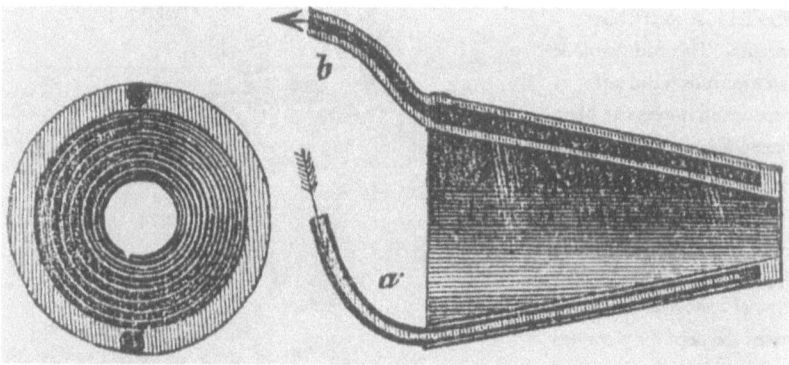

Fig. 2.12. A metallic water-cooled tuyere. An essential element of the equipment of a hot-blast furnace was the water-cooled tuyere. This illustration depicts a water-jacket variety of cooled tuyere. Fresh water ingressed through the feed line (a) and was pumped to the nose, the hottest portion of the tuyere. From this point, the water circulated freely through the hollow jacket of the tuyere, finally egressing through the waste line (b). From Overman, *The Manufacture of Iron*, 419.

loop system.[20] This recycling of the heat was an important step in the economy of the blast furnace and resulted in significant improvements in pig-iron production.

The technical expression of the hot blast was the hot-blast stove, the apparatus in which the blast was heated. The hot-blast stove was usually a brick containment structure, reinforced with an iron framework, into which were vented the hot gases from the furnace tunnel head (top gases). The blast pipe passed through the stove, the fresh blast air being circulated inside a colonnade of arched pipes before egressing the stove. The heat was transferred from waste furnace gas to fresh blast air by convection (fig. 2.13). The hot-blast stove might be situated at the furnace top or at the foot of the stack.

Nineteenth-century ironmaster Frederick Overman recommended first running the waste heat or top gases into the boilers of the steam engine powering the blast machine, then into the hot-blast stove.[21] In either case, some form of *downcomer* pipe was needed either to return the preheated blast air to the tuyeres at the furnace base or to vent the top gases into the boilers of the blast engine. At mid-century, hot-blast stoves raised the blast temperature to about 500 degrees Fahrenheit (260 degrees Celsius).[22]

At first there were some naive applications of the technique of the hot blast; many furnaces simply used specially built fireboxes in which a separate fuel was burned to heat the blast. But the combustion of an independent fuel supply was wasteful in comparison to the alternative of using the hot waste gas vented at the furnace top to preheat the blast. This recycling of the heat that would normally have been expelled as thermal pollution and thereby wasted, and its application in a manner enhancing the rate of production of iron, resulted in a new era of iron smelting.

Iron ores that formerly could not be economically smelted, such as ores with low iron content and ores requiring high reduction temperatures, now were viable stock for the hot-blast furnace. Anthracite coal, which was hard to combust under cold blast, became a popular blast fuel. As noted earlier, this was to change the geography of the iron-smelting industry that prior to 1840 was dominated by charcoal iron production.[23]

All blast furnaces were loaded from the top, but the manner of loading and the character of the structures situated at the furnace top changed during the century. Hand-loading of the furnaces in the first half of the century required that a working platform be created at the furnace top. This structure was the *charging deck*. There were many variations of the charging deck, but all were elevated working platforms where the furnace charges were assembled and loaded into the stack through its open top or through loading doors in the side of the upper stack. The term *stock house* referred to a building on the charging-deck level in which the raw materials were stored (stockpiled) prior to charging. The *keeper* maintained the stockpiles and measured out individual charges.

In the early part of the century, blast furnaces were charged at the tunnel head by wheelbarrows trundled across the charging deck by *top men*, or *fillers*. Some mechanical hoppers were in use in the antebellum period, but most furnaces were charged by manual labor. In the 1870s, mechanical systems supplying the charge to the tunnel head, such as the *skip-hoist*, were widely adopted in the United States. By the late nineteenth century, completely mechanized systems of charging were in use, delivering controlled volumes of ore, flux, and fuel to automatic charging mechanisms at the furnace top, all managed from ground level. Charges had to be spread evenly atop the burden inside the stack, in layers of fuel, ore, then flux. This was done manually until the development in the 1860s of the *bell hopper*, a device that automatically deposited the charges evenly.

The maintenance of an operating supply of fuel, flux, and ore at the furnace plant involved much hand labor in the early part of the century. However, industrial railway systems were developed for materials management as the century progressed. In the early 1800s, raw materials arrived

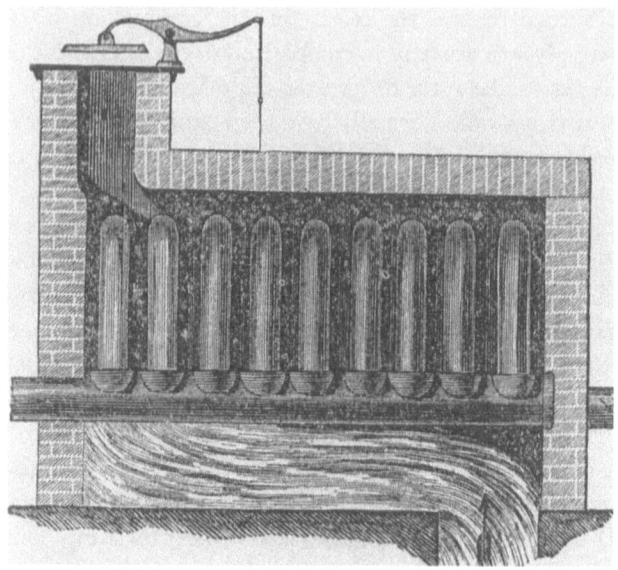

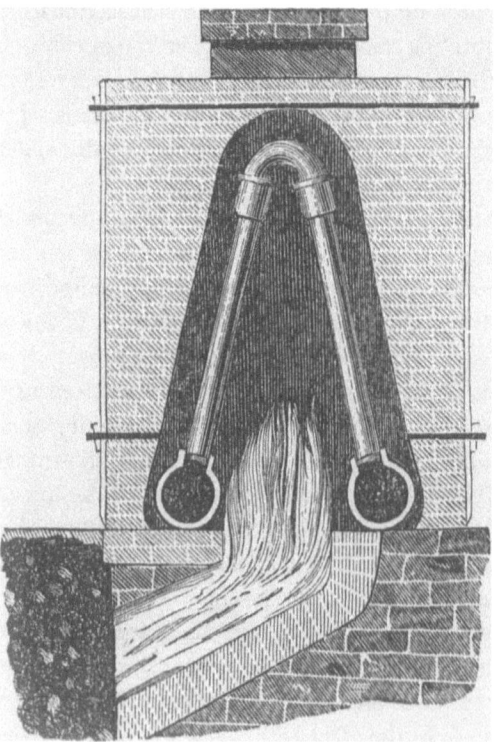

Fig. 2.13. Sectional views of hot-blast stoves. Two separate but similar ovens are shown in profile in these cutaway views. Both were designed to sit atop the blast-furnace stack. In the side view above, hot gases from the furnace top enter the stove from lower right. The gases circulate around the arched pipes and then egress out the short stack covered by an adjustable damper plate. In the end view below, we see the main ingress and egress pipes and the arched connecting pipes through which the pressurized blast is pumped. The fresh, pressurized blast air in the pipes draws heat from the hot waste gases in the masonry shell of the stove. From Overman, *The Manufacture of Iron*, 430, 431.

in wagon loads at furnaces and were unloaded by workers wielding shovels; in the latter part of the century, multiple train-car loads of materials arrived at furnace plants and were unloaded by machinery.

The *casting shed* adjoined the base of the furnace. To protect the casting operation from the elements, the casting shed was normally enclosed, though provisions for adequate ventilation had to be made. The size of the casting shed depended upon the range of products cast from the hearth, but fireproof construction was obviously very desirable; masonry structures thus predominated. In late nineteenth-century practice, enormous coke-fired furnaces casting large volumes of pig iron often did so under the open-sided canopy of a large sheet-metal roof supported by a cast-iron or steel structure.

The Smelting Process

The chemical and thermal reactions taking place inside the blast furnace were not well understood in the nineteenth century. Consequently, smelting was something of an arcane art, and the empirical knowledge of materials and processes accumulated by the ironmaster during his lifetime was perhaps the most important asset of a blast-furnace operation.[24]

The reduction of iron ore in a blast furnace takes place in a thermally insulated structure where heat generated by the combustion of fuel brings about chemical changes in the ore, the process being facilitated by fluxing agents. Materials move in two directions. The burden of the furnace—fuel, ore, and flux—is drawn down by gravity. As the ore breaks down chemically, molten metal flows into the base of the furnace. At the same time, a column of hot gases ascends the stack, heating the burden and causing chemical changes in the raw materials.

The process of smelting iron in the blast furnace can be understood in terms of these two movements: the descent of the burden into the furnace and the ascending column of heat and gases generated at the hearth. In each zone of the stack, the burden undergoes physical and chemical changes. We will follow J. E. Johnson's explanation of the smelting process.[25]

The Descending Raw Materials. Johnson divides the changes the raw materials undergo within the stack into four main categories that correspond to four main zones within the blast furnace: (1) drying and warming; (2) reduction of the ore, decarbonization of the stone, and partial solution of the coke; (3) incipient fusion of the metal and early stages of slag formation; and (4) the final reduction of the ore and separation from the slag.

In the region immediately below the top of the charge, the incoming materials are warmed and dried by the hot ascending gases. In passing through the second zone (the zone of reduction), the ore and the carbon monoxide (CO) produced during combustion of the fuel react to reduce the ore to metallic iron and carbon dioxide (CO_2). As the oxygen of the ore is diminished, the remaining quantities require a higher temperature and carbon monoxide concentration to be released. These conditions are achieved during the slow descent through this, the second and longest stage of the smelting process.

The zone of incipient fusion and preliminary slag formation is located near the top of the boshes, where materials are transformed from a pasty state into a fluid one. The ore begins to melt, forming a siliceous slag with the iron oxide flowing down over the lime and fuel. As the temperature increases, the remaining iron oxide is reduced and replaced in the siliceous mixture by calcium and magnesium oxides from the limestone flux. Near the tuyeres, the coke, ash, and any excess lime combine with the slag.

The highest temperatures exist within a zone between a short distance above and below the tuyeres, the zone of final reduction of the iron. In this area the slag achieves its final composition as the fuel releases its slag-forming matter and the last of the iron is reduced. Carbon absorption is completed as the small streams of iron trickle over the incandescent fuel into the hearth.

The Ascending Gas Column. The gas column in the blast furnace undergoes a complex series of changes in composition, temperature, pressure, and density. The blast entering at the tuyeres produces carbon dioxide as it rises through the burning fuel, then carbon monoxide as it is immediately exposed to excess carbon from the fuel above the tuyere zone. Passing through the charge, the carbon monoxide of the ascending gas relieves the ore of its oxygen, thus converting back to carbon dioxide. The exiting top gases consist primarily of nitrogen, carbon monoxide, and carbon dioxide, with varying amounts of other vaporized elements.

Overman stated that the highest temperatures achieved in the iron smelting furnace (at mid-century) ranged from 2700 to 3000 degrees Fahrenheit (1480 to 1650 degrees Celsius), while the lower limit was set by the fusion temperature of iron.[26] Johnson indicates that, in early twentieth-century coke practice, the issuing iron and slag ranged in temperature between 2650 and 2850 degrees Fahrenheit (1450 and 1570 degrees Celsius),[27] while Overman cites experiments showing furnace slags fused at 2500 to 2600 degrees Fahrenheit (1370 to 1430 degrees Celsius).[28] This, however, was not necessarily the temperature at which it would flow from the hearth. As indicated above, the highest temperatures existed in a narrow zone near the tuyeres where the blast contacted the burning fuel. The temperature decreased quickly above the tuyere plane and then relatively uniformly from the top of the boshes upward, as the endothermic (heat-consuming) iron reduction reaction took place.

The pressure of the ascending gas column was highest in the tuyere zone where the blast entered. Above this point it quickly dropped due to the physical resistance of the charge. The decrease in pressure was fairly uniform between the boshes and the top of the stack. Johnson states that "a uniform and regular pressure drop is both the condition and the product of smooth and regular working".[29] In the absence of manometric (air pressure) data from blast furnaces of the mid-nineteenth century, we can only assume that these general conditions probably held true. Overman indicates that a charcoal blast furnace required one thousand to two thousand cubic feet of air per minute, and a coke furnace three thousand to five thousand cubic feet of air per minute entering at the tuyeres.[30]

Blast-Furnace Operations

A key concern in the management of the blast was consistency in the quality of raw materials and the maintenance of ample stockpiles of these materials. The first consideration was crucial in the early days of furnace operation. Trial-and-error experiments led to the adoption of workable fuel, flux, and ore mixtures. The chemistry involved was not well understood, and the effects of mixing new materials with old ones could not necessarily be predicted in advance. Variations from these practical formulae tended to yield surprising, if not unsatisfactory, results.

After construction of the furnace or its relining with fresh firebrick and refractory sandstone in the hearth, the lining was slowly dried over a period of weeks. When the lining was thoroughly dried, a light load of fuel, ore, and flux was loaded and the blast was applied. The charges and blast pressure were slowly brought up to full over a period of three to four weeks. Once in blast, the operation of the furnace ran around the clock, with tapping occurring about every twelve hours until a production quota had been filled. Any other form of stoppage was the result of some accident or oversight. The interaction of the furnace plant, the work force, and the raw materials was continuous.

The furnace was charged constantly by the fillers, the usual practice being to load first fuel, then ore, and then flux in alternating layers, the amounts depending on the quality and characteristics of each. A uniform level of burden was maintained in the stack; this was the *stock line*. The quantities of fuel, ore, and flux charged into the stack varied depending on the chemical qualities of the raw materials. Ure cites two examples of charges loaded into large British coke furnaces in one day. In the first example, 36,000 pounds of coke, 36,000 pounds of iron ore, and 18,000 pounds of flux were loaded in sixty charges.[31] In another example, 28,660 pounds of coke, 32,000 pounds of ore, and 13,500 pounds of flux were charged over twenty-four hours, a total of 74,160 pounds of raw materials yielding seven tons or 14,000 pounds of pig iron.[32] Viewing these figures, it is easy to appreciate the enormity of the problem of obtaining raw materials and maintaining stockpiles at the furnace.

At the furnace base were workers adjusting and cleaning the tuyeres, pulling slag, and tapping the iron—all under the direction of the *foreman* or a *founder*. During the blast, examination of the slag was the most reliable indicator of the functioning of the furnace. According to oral tradition, a dark green color indicated an optimum blast while black slag was thought to be a sign of an inefficient furnace.[33] As the reservoir of the lower hearth filled with molten iron, the time for tapping the hearth approached. The liquid slag, carrying with it impurities from the molten metal, floated on top of the denser molten iron. As the iron filled the hearth, the suspended slag would flow over the damstone. In late nineteenth-century furnaces, a special *cinder notch* vented the slag from the hearth.

The slag flowed and was drawn off the top of the molten iron by a worker stationed at the mouth of the hearth. At the time of tapping, the blast was suspended and the clay plug in the damstone was broken, releasing the heavy metal in a stream down a runner trough laid in the sand floor of the casting shed. Once tapped, the damstone was again plugged with fire clay, the blast was brought up to pressure, and the charging resumed.

The smelting of iron ore resulted in two major by-products. The first was heat. By the middle of the nineteenth century the recycling of hot gases from the tunnel head through the hot-blast oven and blast generators (steam engines) had begun in North America. The second by-product was slag, a hard, glassy substance containing the ore gangue and other mineral wastes from the smelting process. After the blast, there were few uses for the slag, at least until the late 1800s.[34]

Molten iron tapped from the hearth could either be cast as pig iron or direct-casted into consumer items. Pig iron was not a finished product per se; it merely represented the primary reduction of ore into a usable metallic form. The pig iron was then sold to foundries, forges, and rolling mills for further processing. However, at many furnaces, particularly in the antebellum period, the molten iron was cast directly into usable items such as fireplace backs, kettles and pots, and stove plates. In this event, the furnace employed the services of a *foundryman* and *mold maker*.

The floor of the casting shed was covered with a deep layer of carefully selected fine sand. When lightly wetted, the sand retained the impressions of wooden molds pressed into it. Pig iron (and other flat objects such as stove plates) were cast from the furnace directly into impressions in the floor of the casting shed. The iron tapped out of the hearth flowed by gravity into the casting area following troughs molded into the shed floor. As iron flowed into the casting shed, its course was directed by *guttermen*. Like a field of crops irrigated by a network of ditches, the iron was successively turned off the main stream into smaller channels running at right angles. Off these feeder channels or "sows" were the individual impressions or cavities for bars of iron called "pigs."

When the iron chilled, it formed a solid, brittle mass. The pig-iron bars were then broken off the sow runners using heavy sledgehammers. A typical pig-iron bar might measure three feet in length, be four inches in width and depth, and weigh nearly 250 pounds. The feeder channel iron or sow was broken up and carted off, as were the broken fragments of the main furrow or main runner. Pig-iron bars and the miscellaneous scrap fragments were sold by weight.

Many finished consumer items were cast in the shed, using molten metal tapped directly from the blast-furnace hearth. Molders and patternmakers constructed wooden models of the items to be cast and also boxes or flasks which were then filled with casting sand. The wooden model was pressed into the sand, leaving an impression. With the wooden mold removed, the cavity was then filled with molten iron. For these individual boxed molds, the iron was tapped out of the furnace into a hand-held ladle or a larger suspended crucible. The ladle or crucible was then carried by the *ladleman* to the molds, and each impression was carefully filled with iron. Complex forms required detailed molds with air vents and casting slots known as *gates* and *risers*. Large items could be cast in the floor of the casting shed in the same manner as pigs. As the century progressed, blast furnaces specialized more and more on the production of pig iron alone, leaving the final casting of complex forms to the foundry.

The furnace worker's job was a grueling one for most of the century; once in blast, the operation was nonstop. The daily shift was twelve hours, seven days a week. Not until the twentieth century did organized labor unions effect changes in this basic work schedule.[35] The long-term effects of the environmental hazards on workers' health were not considered.

Difficulties were bound to arise in an environment inhabited by people, materials, and machines, all of which displayed individual quirks, weaknesses, and unknowns. One of the truisms about the furnace in blast is that one small failure could bring the entire operation to a screeching halt. Fatalities and often horrendous injuries could occur when things went wrong. While a host of blast problems were known, solutions were sometimes lacking.

A common failure during smelting was a loss of blast; without the blast to push the fuel combustion, the burden of the furnace in the hearth area would chill into a solid mass. The loss of blast might be caused by failure of the motive force driving the blast machine (the steam engine or waterwheel/turbine), a failure of the blast machine proper, a rupture in the blast pipe causing a loss of full or partial blast pressure, or the clogging or disintegration of a blast nozzle and tuyere. The cooling of the tuyeres in a hot-blast furnace was a critical function. Without the cooling, the tuyere would disintegrate and the nozzle would fail. Volumes of water introduced into the hearth of the furnace through the ruptured tuyere cooled the burden and created hazardous steam.

Scaffolds were congealed masses of fused ore, flux, and fuel adhering to the wall of the furnace lining. Scaffolds were created when cold spots occurred in the burden, typically due to failure of one of the tuyeres or as the result of the poor distribution of fuel in the charge. As the burden chilled, the pasty metal-and-fuel mix congealed and adhered to the lining, causing a shelf to form. The resulting scaffold impeded the downward progress of the burden, causing a "backup." Often the scaffold would go undetected, and as the burden below the scaffold progressed downward, a cavity formed beneath the shelf. This cavity then filled with explosive gases. If the scaffold gave way suddenly, the burden would descend in an erratic, abrupt manner, resulting in equally abrupt strains on the blast equipment and disturbing the equilibrium of the smelting. Explosions in the gas cavity might also result.

If the charges loaded at the furnace top were disproportionate in any way, the consequences would not necessarily be noticed until it was too late to do anything. After all, the furnace was charged continually, and a bad charge took many hours to works its way to the hearth. From the top of the stack, there was no way to alter what was happening at the hearth level. An insufficiency of fuel in the burden might result in a mass of partially reduced metal congealing in the hearth.

If the burden chilled in the furnace, the result was a *salamander*, a conglomeration of reduced iron, partially reduced iron, partially combusted fuel, and incompletely formed slag. With luck, the furnace hearth would have been tapped before the cooling occurred. Otherwise, the base of the salamander might consist of several tons of pure iron fused to the sandstone hearth lining. As the size of furnaces increased during the century, the size of the potential salamander grew proportionately. Frequently, in the late nineteenth century, furnacemen had to resort to the use of dynamite to break up the salamander into pieces that could then be cajoled out of the hearth. Once the salamander was removed, the hearth and stack had to be relined up to the boshes.

Furnacemen evolved many tricks of the trade to correct problems occurring during a blast, but there was always an element of physical danger in operating the furnace. A more subtle threat was economic; given the enormous investments required to run them, furnaces that operated inefficiently or produced undesirable grades of iron had a limited lifetime.

The Industrial Milieu

Early furnace operations were often isolated from population centers, resulting in the creation of "iron plantations." Workers not only mined ore, charred wood, and produced iron but furnished their own food supply and building materials and fashioned their own settlements. Large corporate enterprises were sited near population centers, although the workmen might well live in a company-built suburb near the plant. As furnace companies increased in size during the century and integrated with other iron manufacturing enterprises, the iron plantation vanished.

Pig iron produced at the blast furnace was reworked into two principal forms: cast or wrought iron. The pig iron produced by the blast furnace was a form of cast iron consisting of metallic iron homogeneously mixed with impurities that gave the iron strength but also brittleness and hardness.

This type of iron was suitable for many but not all ferrous metal applications. At the *foundry*, pig iron was remelted in metal cupola furnaces and cast into myriad forms. Machine shops took the rough casting and, depending on the characteristics of the remelted iron, manufactured the casting into a finished product by lathing, planing, drilling, grinding, and polishing. Malleable cast iron could be tooled, but simple gray cast iron could not. Malleable cast iron combined the fluidity (during casting) of cast iron and the ductility of steel. Some cast items produced in the foundry, such as cutlery and harness hardware, were exposed to high heat in an annealing furnace for extended periods of time. The combination of carbon from the casting with oxygen from combusting fuel produced carbon monoxide, which was expelled as a gas. The result was crucible steel, an expensive commodity in the early nineteenth century.

Wrought iron is really cast iron that has been made malleable by the removal of specific impurities, including carbon, silicon, sulfur, and phosphorus. Its physical structure is a mixture of more-or-less pure fibrous iron with impurities and slag situated in strands between the iron crystals. The physical and chemical structure of wrought iron made it soft and malleable when heated. To make wrought iron, pig iron was remelted in a *refinery forge* or, in the late nineteenth century, a *puddling* or *reverberatory furnace*. A lump or puddle of pasty metal known as a *bloom* was removed from the puddling apparatus and then mechanically pounded and shaped under a forge into a bar. This bar was not unlike a bar of pig iron, for it was an intermediate step toward transformation into a consumer item.

At the *rolling mill*, wrought-iron bars still hot from the puddling and forging process were passed through a series of compressing and shaping rollers. The end products were long rods, bars, heavy plates, angle irons, and thin sheets of strong but flexible iron. One of the most important types of rolling mill in nineteenth-century America was the rail mill. Railroad iron (rail) could be produced only at the largest of plants. Rail mills were the rarest of the iron-fabricating plants, in part due to the speciality of the trade and to the high cost of the machinery needed to turn out long lengths of a heavy wrought-iron product.

Research into blast-furnace technology at the middle of the nineteenth century provided us with a basic vocabulary with which to explore historical documents of the period. The research also gave us a set of general expectations as to what we might find buried at the Bluff Furnace site and a technical background for identifying what was found.

Of the two lines of inquiry we pursued, namely history and archaeology, the documentary evidence was the easier from which to extract a story. Several secondary sources, specifically histories of iron manufacturing, had provided the seminal narrative of the events at Bluff Furnace. To reconstruct the events at the site in greater depth, as well as provide the historical and industrial context of the operation, we searched both primary historical documents—those made in the mid-1800s by persons associated with Bluff Furnace—and secondary sources that discussed nineteenth-century iron manufacturing in general.

CHAPTER 3

Bluff Furnace and the East Tennessee Iron Manufacturing Company

Iron manufacturing is rooted not only in dispassionate geological processes but in the passionate accidents of human history. The artificial geopolitical boundaries of Tennessee and its counties subdivide an ancient landscape arranged in its own natural order. The Great Valley of East Tennessee, a fertile, well-watered valley comprising the drainage of the Tennessee River and its principal tributaries, the French Broad, Holston, and Powell rivers, forms an avenue some twenty miles wide and two hundred miles long, extending from Bristol, Virginia, to Chattanooga, Tennessee.

The Great Valley is framed on the east by the Unaka Mountains, the southern division of the Appalachian chain, which extends from New York and Pennsylvania southeast across Virginia, Kentucky, and Tennessee into northern Alabama. On the west, the valley is bounded by the escarpment of the Cumberland Plateau, a high, flat plain marking the western extent of the geographic unit known as East Tennessee. Near the southern end of the valley, where the states of Georgia, Alabama, and Tennessee meet, the Tennessee River begins a tortured, winding passage through the heights of the Cumberland Plateau.

Three areas of the state contained iron-ore deposits exploitable on a commercial scale, and two of these regions were situated in the Great Valley. The *Eastern Iron Belt* was a mineral region containing hematite, limonite, and magnetite deposits situated along the western edge of the Unaka Mountains. Of the three types of iron ore in the belt, limonite was the most common, and magnetite the least. The *Dyestone Belt* consisted of hematite ore deposits along the eastern edge of the Cumberland Plateau. Appearing as seams in surfaced outcrops of buckled ridges and in the exposed flanks of the Cumberland Escarpment, the ore banks were seldom more than a few feet thick and often stretched in thin, broken strands across a dozen counties. The *Western Iron Belt* of Middle Tennessee consisted of limonite ore deposits scattered along the length of the physiographic province known as the Western Highland Rim, the elevated landform west of the basin in which Nashville was to grow. Geology allied this iron region with that of western Kentucky, and the exploitation of the Western Iron Belt had little in common with ore smelting in East Tennessee.

Only one commercial coal region was present in Tennessee; substantial bituminous coal deposits were present in numerous strata forming the Cumberland Plateau. Depending on the particular stratum of coal, these deposits were mined in shallow pits, in shallow cuts into the flanks of seams exposed in escarpments and erosional gulfs, or by drift mines and tunneling. The coal fields extended into northern Alabama and central Kentucky.[1]

Of the three iron-producing regions of the state, the Dyestone Belt was perceived in the nineteenth century as holding the greatest economic promise because of the close proximity of hematite iron ore, bituminous coal from the Cumberland Plateau, and an abundance throughout the region of limestones suitable for smelting fluxes. On the other hand, ore and coal deposits within the Dyestone Belt were thin and dispersed, and proved difficult to mine on a large commercial scale in the early 1800s. This factor played an important role in the historical development of the state's iron industry.

Antebellum Ironworks of Tennessee

The valley of the Tennessee River was the path through which white settlers in the late eighteenth and early nineteenth century funneled southwest into the heartlands of the Cherokee and the Creek Indians. Deposits of coal and iron in East Tennessee were exploited by European-American pioneers from the earliest period of settlement. The numerous small veins of iron ore were easily tapped by small frontier ironworks. Abundant forests provided a seemingly endless supply of charcoal, and the profusion of fluxing limestones and dolomite contributed to the steady growth of Tennessee's early iron industry. The scale of these early enterprises was small, in keeping both with the nature of the mineral resources and with the sparse frontier population of the region.[2]

While landforms such as the Unaka Mountains and the Cumberland Escarpment contained exploitable mineral resources, they also presented substantial barriers to population movement and economic development in the late eighteenth and early nineteenth centuries. Like the communities they served, the early furnaces and forges of East Tennessee were numerous but small and commercially isolated from large population centers and markets. The products of the remote ironworks served an agrarian frontier community, providing cast hollowware (kettles, pots, and pans), stove plates, and firebacks, and wrought-iron bars for nails, tools, wagon tires, mill gear shafts, and agricultural implements.

More numerous than blast furnaces in the hills of East Tennessee were primitive bloomary forges where ore was smelted in small charcoal-fed furnaces and then pounded by water-powered trip hammers into wrought-iron bars. Unlike the brittle cast iron from the blast-furnace hearth, the malleable wrought-iron ingots could be hammered, cut, split, and welded by blacksmiths into innumerable items including nails, horseshoes, door hinges, wagon tires, and axe heads. On the Tennessee frontier, bars of wrought iron were a valuable exchange commodity and were frequently used as a form of currency.[3]

The development of iron-smelting facilities followed soon after white settlement. During the territorial days when East Tennessee was a trans-Allegheny province of North Carolina, David Ross, an experienced Virginia ironmaster, obtained bounty lands for the development of iron manufactures and, in 1789 or 1790, erected Tennessee's first ironworks on the North Fork of the Holston River in Hawkins County. After entering the Union in 1796, the ever-expanding state of Tennessee promoted the establishment of iron industries with legislation.

General John Sevier, a prominent military commander during the American Revolution, entered into a partnership with one Walter King and commenced the erection of an ironworks in Sullivan County. About 1811, these ironworks on Kendricks Creek were taken over by brothers Elihu and Elijah Embree and given the name Pactolus Ironworks. The works at Pactolus were large: in addition to producing pig iron, the Pactolus forges produced wrought iron. The output of the nail works, in particular, was widely known.[4]

Middle Tennessee's first furnace was erected in 1795 on the Iron Fork of Barton's Creek by James Robertson. Cumberland Furnace, as it was called, was sold to Tennessee's most famous antebellum ironmaster, Montgomery Bell, in 1804. Bell became involved in an ambitious scheme to develop a major industrial site at the Narrows of Harpeth on the Harpeth River. In

1850, Bell's iron plantation at Patterson Forge employed a work force of slaves numbering 322 and was one of the largest iron concerns in the South.[5]

Exploitation of iron resources in lower East Tennessee first required an accommodation between the European-American and Native-American populations. The mineral resources of the Chattanooga area were recognized at an early date, decades before the Cherokee inhabitants of the area were vanquished. With the support of the U.S. government, Colonel Elias Earle of Greeneville, South Carolina, explored the Cherokee Nation in 1807 with a view toward establishing an ironworks. Earle selected a six-square-mile tract on the south bank of the Tennessee River at the mouth of South Chickamauga Creek. A tentative agreement to purchase the tract from the Cherokees was made, and a crew of workmen to build the furnace was dispatched to the site. The crew was intercepted near present-day Ringgold, Georgia, by a group of armed Cherokees hostile to the incursion. The workmen were detained for several days and released only after being warned not to return to the Cherokee Nation. The iron-making enterprise dissolved and was not revived.[6]

Purchase of the Hiwassee District above Chattanooga from the Cherokee in 1819 opened the area north of the Tennessee River to European-American settlement. Hamilton County was formed in that year, its county seat being Dallas, a dozen miles northeast of Chattanooga on the north bank of the river. Agitation to remove the Cherokees beyond the Mississippi River increased, the prize being rich agricultural lands along the drainage of the lower Tennessee River. Surrendering to increasing pressure, the Cherokee people ceded their remaining territory south of the Tennessee River to the United States in the 1835 Treaty of New Echota and abandoned the area under duress in 1838. As the Cherokees retreated from the Great Valley, the white Americans were close behind, exploiting the mineral resources untouched by the land's aboriginal inhabitants.

Bluff Furnace ironmaster Robert Cravens began his life at one end of the Great Valley, and when it was spent, he found himself at the other end. He was born in Rockingham County, Virginia, in 1805 and spent his youth in Greene County in upper East Tennessee. His parents, James and Ann Love Cravens, moved the family to Selma, Alabama, in 1821, but within a short time, a fever claimed both their lives. Young Robert, now in charge of his surviving seven siblings, returned to the familiar surroundings of Greene County and the support of his extended family. Robert's maternal aunt, Mary Love Gordon, was married to ironmaster George Gordon, who operated Bright Hope Furnace. The Gordons were childless and took the Cravens children under their wings. At age sixteen, Robert entered the iron business under his uncle's tutelage.[7]

Robert Cravens moved from Greene County one hundred miles down the Tennessee River to Rhea County about 1828 and in 1831 married Catherine Roddy. Pioneer ironmaster Mathew English had established a bloomary forge on White's Creek prior to statehood and in the early 1820s went into partnership with George Gordon, constructing a new furnace at the old bloomary site.[8] Upon English's death, Robert Cravens bought a half interest in the operation. Cravens began a family and familiarized himself with iron smelting in the upper end of the Dyestone Belt.

By the end of the 1830s, Cravens was in partnership with Jesse Lincoln, a cousin of Abraham Lincoln. Lincoln's industrial capital consisted mainly of slaves; in 1838, this work force began construction of a blast-furnace complex off the Tennessee River near the mouth of White's Creek at the southern tip of Roane County. Eagle Furnace was completed in 1839, and the iron plantation included a flouring mill, workshops, and a foundry. A canal was dug to the nearby Tennessee River to facilitate barge shipment of heavy iron products to markets downriver.[9]

Eagle Furnace was thirty-three feet high, eight feet across at the boshes, and burned charcoal to smelt the local red hematite or dyestone iron ores.[10] Eagle Furnace performed well, the pig iron being barged downriver to the Mississippi for transhipment to Cincinnati and St. Louis, while mule teams carted oven plates, pots, and other finished castings on treacherous roads to regional markets in Tennessee, Kentucky, and Georgia.[11]

In 1844, Cravens built a brick blast furnace for experimental purposes. Called "Eagle Furnace No. 2," this charcoal-fired, cold-blast furnace was only twenty feet tall and four feet across at the boshes. It was abandoned after a short blast of five or six weeks, having produced slightly more than a ton a day of "very poor iron."[12] Just what was experimental about the furnace, other than its small size, is not known. This business failure was followed a year later by the death of Cravens's wife, Catherine, the mother of his five children. Robert was remarried in December 1846 to Caroline Cunningham but had no more children.

Despite the tragedies in his personal life, Cravens's business continued to prosper. He purchased Lincoln's interest in the White's Creek operation in April 1845 for the sizable sum of twelve thousand dollars. Included in the conveyance were the furnace, forge, and mills, as well as real estate, livestock, and nine slaves.[13] A bloomary forge was added to the Eagle Furnace complex in 1848 but was abandoned by 1850. The water power used to operate the forge was absorbed by a new grist and saw mill. The enterprise at White's Creek was now a full-fledged iron plantation, a small, self-sufficient industrial community. Eagle Bloomary Forge No. 2 was built in 1855 and consisted of one bloomary fire and one water-driven hammer.[14]

In 1848, Robert Cravens engaged in a risky and ultimately unsuccessful attempt to utilize at Eagle Furnace the latest blast-furnace technology then in use in the United States. Edward E. Andreae was hired by Cravens to convert Eagle Furnace to a hot-blast, coke-fired plant, a configuration at that time in use in America at only a handful of furnaces in the Northeast. The conversion failed, and Andreae filed suit against Cravens for unpaid wages.

To compare charcoal fuel consumption with that of coke, Cravens had insisted that the furnace blast begin with charcoal and then switch to coke. Consequently, Andreae built a hearth suitable to burn charcoal. Andreae desired to build the hearth fifteen or eighteen inches higher for the coke trial but surrendered to Cravens's demand to start the blast with charcoal. A hot-blast apparatus had also been built, and in the course of the fuel test a preheated blast would be introduced and changes in fuel economy noted. As part of the test of the new fuel, the remelting of pig iron in the foundry cupola would also be done with coke.

The deposition of the single witness in the case revealed that Andreae successfully fired the foundry cupola at White's Creek with coke, although the iron produced was said by some of the molders to be too brittle for some purposes.[15] None of the documents, however, indicate that the planned coke-fired blast of the main furnace was ever actually carried out.

Andreae had been financially destitute at the time of the proceedings and, as a pauper, had received the services of court-appointed attorneys Thomas C. Lyon and Albert G. Welcker. Interestingly enough, Lyon was one of Robert Cravens's partners in the East Tennessee Iron Manufacturing Company and, as of July 1850, had a financial interest in Eagle Furnace. When Andreae had arrived at Eagle Furnace in September 1848, Cravens had already launched his most ambitious industrial venture.

The East Tennessee Iron Manufacturing Company

Cravens and other iron manufacturers in East Tennessee found themselves in an economic situation akin to straddling a narrow rail fence. On the one hand, there was a small, steady demand for cast- and wrought-iron products in local markets, but on the other hand, only limited access to national markets. Large, integrated iron plants, such as those in the North, could produce more cheaply, volume production permitting volume pricing—the economy of scale.

In the Great Valley of East Tennessee, iron manufacturers were prisoners of their restricted market access. Road networks were slow to progress through the spectacular but treacherous mountain scenery of the region. The Tennessee River provided a connection with the Mississippi and its attendant ports, but the river was a less-than-ideal business partner. Three months out of the year, the water levels were too low to permit passage over a number of shoals and obstructions in the channels. At the other extreme, devastating floods were just as common. Commercially, the Tennessee River was a "severed artery," being split at the fourteen-mile-long river obstructions at Muscle Shoals into upper and lower rivers. At this point, most steamboat cargoes had to be portaged overland around the dangerous shoals, although shallow-draft barges could be floated past.

Railroads offered a solution to this bottleneck. Two railroad lines played important roles in the settlement and industrialization of Chattanooga and lower East Tennessee: the Western and Atlantic (W&A) Railroad, a state railway chartered by Georgia, and the Nashville and Chattanooga (N&C) Railroad, chartered by the state of Tennessee. The trading post known as Ross's Landing was renamed Chattanooga in 1839 after its inhabitants learned that their small hamlet had been selected as the northern terminus of the W&A Railroad, one of Georgia's three state-owned railways chartered to open internal rail transport in the state. The line, which stretched 137 miles from Atlanta to Chattanooga, was built in fits and starts, beginning in 1836. The route was serviceably completed to Chattanooga in May 1850.

The N&C originated at the state capital and coursed southeast toward a crossing of the Tennessee River at Bridgeport, Alabama. The railroad then turned north and threaded its way through the Cumberland Plateau, hugging the cliffs above the river and entering Hamilton County at

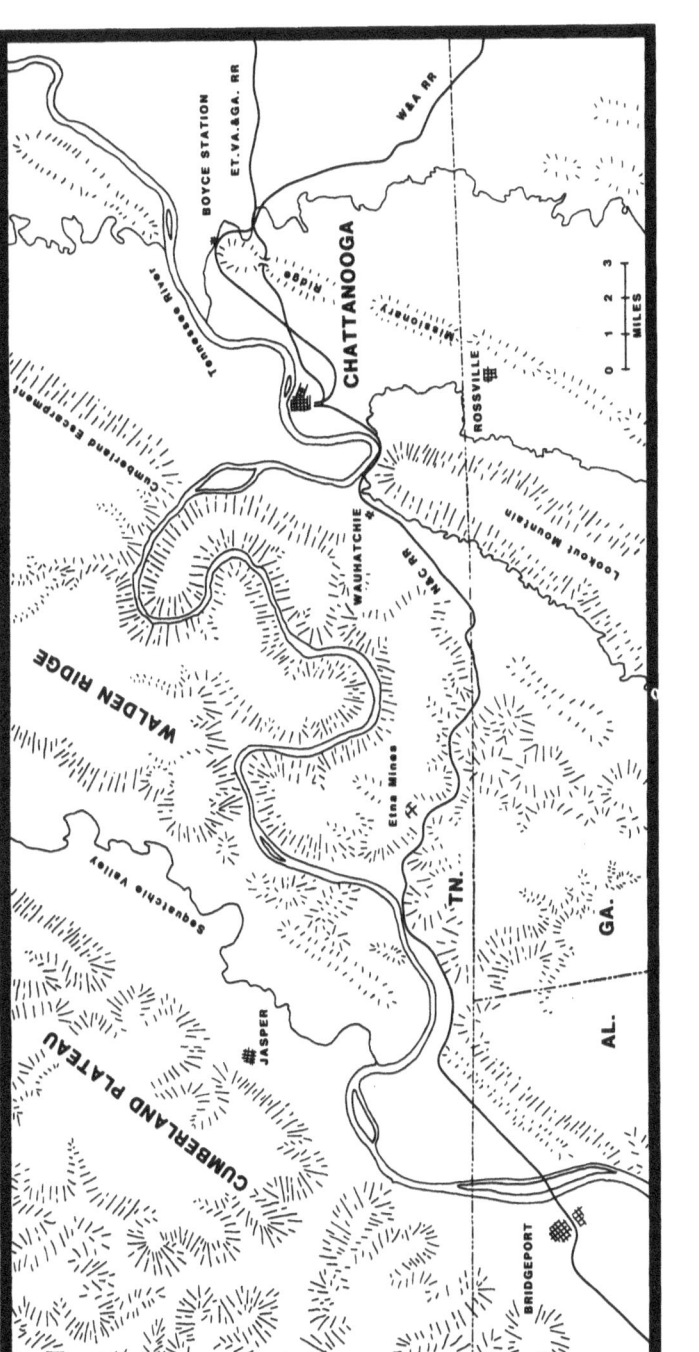

Fig. 3.1. Chattanooga and the Tennessee River Valley in 1860.

Fig. 3.2. a. and b. Robert Cravens and James A. Whiteside. The rugged visage of ironmaster Cravens, at left, contrasts with the polish of lawyer, politician, and entrepreneur Whiteside, at right. Cravens portrait courtesy Cravens House, Chickamauga and Chattanooga National Military Park, National Park Service. Whiteside portrait from an engraving in Hale and Merritt, *A History of Tennessee and Tennesseans*, vol. 6.

Fig. 3.2 a.

the foot of Lookout Mountain. When the N&C route was completed into the town in 1854, Chattanooga had acquired its major rail arteries (fig. 3.1).

The Memphis and Charleston Railroad opened its line in 1858, connecting Tennessee's largest and commercially most important city with its seaport counterpart on the Atlantic coast of South Carolina. Still later, the main line of the East Tennessee, Virginia and Georgia Railroad from Knoxville terminated on the W&A line at Dalton, Georgia, and in 1858 was connected by spur directly with Chattanooga.

By 1860 Chattanooga was a regional railroad hub with connections to most of the principal cities of the Southeast, including Charleston, South Carolina; Atlanta, Georgia; Nashville, Knoxville, and Memphis, Tennessee; and Bristol and Richmond, Virginia.[16] The first W&A trains entering Chattanooga heralded the beginning of industrialization in the region, and the gaze of many an entrepreneur turned toward the hamlet at Ross's Landing as a site for lucrative investments.[17]

For Robert Cravens and the East Tennessee Iron Manufacturing Company, Chattanooga's role as a rail junction was important in two ways. First, the railroad was an industrial tool that permitted reliable, rapid bulk transport of raw materials and the shipment of finished products to several major markets in the Southeast. Second, the railroads themselves consumed enormous quantities of wrought and cast iron in the form of

Fig. 3.2 b.

rails, track fixtures, and spikes, along with literally hundreds of metal parts such as axles, wheels, and drawbars, for rolling stock and locomotives. Construction and maintenance of rolling stock and track required the services of machine shops, foundries, and rolling mills. Serving the railroads and being served by their presence expanded the industrial future of iron manufacturing in the Great Valley.

Robert Cravens and other investors in the valley laid the groundwork for their new enterprise in iron. The East Tennessee Iron Manufacturing Company was given a corporate charter by an act of the Tennessee General Assembly on November 27, 1847.[18] The company was established "for the purpose of manufacturing iron, machinery and implements, and all articles composed in whole or in part of iron, steel and wood." Officials of the company were to consist of an elected board of five directors, from which a president would be elected. Opening books for subscription to the capital stock, initially limited to a maximum of $250,000, were William Williams, William Swan, Samuel B. Boyd, Thomas C. Lyon, and Robert Cravens.

William Williams and his brother James had grown wealthy on the river commerce of the region and had created Chattanooga's first bank.[19] William G. Swan was attorney general of Tennessee from 1851 to 1854, mayor of Knoxville from 1855 to 1856, and one of the early industrialists of that city.[20] Samuel B. Boyd had a similar background, serving as mayor of Knoxville from 1847 to 1851. Thomas C. Lyon was a member of the Tennessee Bar and frequently served as a special judge of the Tennessee Supreme Court.[21] Robert Cravens was the only practical ironmaster in the initial group of incorporators; the remainder were simply investors.

The five stock subscribers of the corporation were soon joined by others with even more formidable credentials as investors and industrialists. Of all the participants in the East Tennessee Iron Manufacturing Company, James Anderson Whiteside was probably responsible for steering its course as a legal entity in the years prior to the Civil War. Whiteside was a lawyer and a member of the Tennessee legislature who had moved from Pikeville, Tennessee, to settle in Chattanooga soon after the Cherokee removal. He successfully promoted the selection of Ross's Landing as the terminus of the W&A Railroad and spearheaded the incorporation of the N&C. Whiteside returned to the state legislature for the years 1845–47 and was a partner in several land syndicates investing in lower East Tennessee acreage.[22] James A. Whiteside served as president of the East Tennessee Iron Manufacturing Company until his death in 1861.

Ironmaster Robert Cravens was perhaps the corporate counterpart to Whiteside, providing the practical experience needed to successfully operate an iron business (fig. 3.2). Not only did Cravens bring to the company his experience in the trade, he also brought material assets forming the core of the enterprise. On July 6, 1850, in exchange for an unspecified number of shares in the company, Robert Cravens conveyed to the East Tennessee Iron Manufacturing Company his title to "Eagle Furnace, saw and grist mill, forge and workshops," situated on White's Creek in Roane County.[23] An ore bank situated near White's Creek Island was included in the one-thousand-acre conveyance. Cravens thus vested in the company title to a working blast furnace and forge—at that date, the company's only real producing asset. Robert Cravens usually conducted real estate transactions on the company's behalf, signing as "Agent" and occasionally as "Superintendent" on land deeds. Upon the death of Whiteside in 1861, Cravens assumed presidency of the company.

Other members of the company "bought in" by exchanging land for stock. In January 1852, Farish Carter relinquished title to several parcels of land in the district in exchange for fifty thousand dollars' worth of stock shares in the East Tennessee Iron Manufacturing Company.[24] Carter was a prominent upcountry Georgia planter and land speculator from

Murray County.[25] Carter's trade of real estate for stock placed in the company's name substantial assets that would over the next decade be used to provide a continual source of income. The real estate conveyed included land in Hamilton and Marion counties, an interest in Lowry's Island[26] in the Tennessee River, half of a five-thousand-acre tract on top of Walden's Ridge, and portions of fourteen town lots in Chattanooga.

Other stockholders were less well known. Spencer C. Rogers may have been a director or perhaps secretary or treasurer of the company.[27] Rogers (with Robert Cravens) participated in the incorporation of Chattanooga's first, albeit unsuccessful, municipal water corporation, the Chattanooga Waterworks Company, chartered in 1856. The list of stockholders in the iron company also included Samuel Johnson of Roane County and Luke Lee of Chattanooga.[28] Johnson conveyed to the company five hundred acres on the Tennessee River near the Eagle Furnace complex in exchange for twenty shares of stock. The provisions of the sale included privileges allowing the company access to the river and the right to land boats. Further: "Should Cravins [sic] or his assigns erect a Blast Furnace on the same he or they shall have the privilege of the use of such ground adjoining the furnace and over the line as may be needed for wasting slag, stacking metal and castings and such necessary room as may be wanted in connexion [sic] with the Furnace."[29]

At least one of the shareholders in the company wielded influence on a regional scale. Among the most prominent industrialists in the antebellum South was Ker Boyce of Charleston, South Carolina. Boyce sat on the boards of a dozen industrial enterprises and amassed a sizable fortune in the process.[30] His son, James Pettigru Boyce, also held stock in the company, and would inherit the bulk of his father's estate, including shares in the East Tennessee Iron Manufacturing Company.[31]

Cravens and his partners shared a common view of a broadening industrial horizon, but there was as much to risk as to gain. At the middle of the nineteenth century, the iron industry of Tennessee had reached its maturity. The modern geographical limits of the state had finally been reached with the Cherokee removal in 1838, and the incorporation and growth of Chattanooga marked the emergence of what would be the last of the state's four largest cities. On the state's southern border, at the break in the Cumberland Plateau where water and rail transport lines converged, Chattanooga stood as the young commercial lion of Tennessee.

Tennessee charcoal iron had already acquired a national reputation for its quality in foundry castings, particularly for use in railroad car wheels. The pig-iron tonnage produced in the state grew steadily in the 1840s and 1850s but had begun to compress in the years prior to the Civil War. Although largely due to a withering competition with larger, more efficient northern furnaces, the Tennessee furnace operations also suffered from the effects of the economic panic of 1857, a nationwide financial crisis triggered by the failure of a large trust company due to overinvestment and overspeculation in industrial enterprises, particularly railroads.

While Tennessee produced increasing amounts of pig iron in the antebellum period, its contribution to the national total decreased. Many of the small furnaces and bloomary forges erected during the late eighteenth and early nineteenth century were being replaced by larger, more commercially competitive enterprises such as those constructed by Montgomery Bell in Middle Tennessee.

In 1840, there were thirty-four furnaces and a total of ninety-nine bloomaries, forges, and rolling mills in Tennessee; the furnaces produced 16,129 tons of castings (including pig iron) while the latter group produced 9,673 tons of wrought-iron bars. This represented 5.6 percent of the national total of cast iron and 4.9 percent of wrought-iron bar.[32] Statistics from the U.S. census in 1850 revealed that, among the southern states, Tennessee led in the production of pig iron (table 3.1).[33] Tennessee's production of 30,420 tons of metal, however, accounted for only 5.4 percent of the total output of the United States. State statistics from 1854 document the production of 38,773 tons of pig iron and an additional 1,433 tons of direct castings. In 1860, Tennessee again led the southern states in pig-iron output, but with only 22,302 tons, a sharp decrease from its production levels in 1850 and 1854.[34]

Within Tennessee there were equally striking variations in pig-iron production. Dwarfing the iron industry of East Tennessee was the furnace activity on the Western Highland Rim of Middle Tennessee. In 1831, state geologist Gerard Troost counted only six operating blast furnaces in the region. By 1835, the furnaces numbered twenty-seven.[35] Bloomary and refinery forges were part of most of the furnace enterprises, and several furnaces were integrated with rolling mills. A state geological report by Safford in 1854 put the number of operating furnaces on the Highland Rim at thirty-one, producing in that year a total of 37,183 tons of pig metal and castings: 95.9 percent of the state total for that year.[36] This boom in furnace activity on the Western Highland Rim was short-lived, however, and iron production in the region had begun to

Table 3.1

Pig-Iron Production in the United States, 1850, by State

State	Pig Iron, in Tons	% of U.S. Total	No. of Firms
Alabama	522	.09	3
Connecticut	13,420	2.38	13
Georgia	900	.16	3
Illinois	2,700	.48	2
Indiana	1,850	.33	2
Kentucky	24,245	4.30	21
Maine	1,484	.26	1
Maryland	43,641	7.74	18
Massachusetts	12,287	2.18	6
Michigan	660	.12	1
Missouri	19,250	3.43	5
New Hampshire	200	.04	1
New Jersey	24,031	4.26	10
New York	23,022	4.08	18
North Carolina	400	.07	2
Ohio	52,658	9.34	35
Pennsylvania	285,702	50.68	180
Tennessee	30,420	5.40	23
Vermont	3,200	.57	3
Virginia	22,163	3.93	29
Wisconsin	1,000	.18	1
Total	563,755	100.02%	377

Source: Debow, *A Statistical View of the United States* (1854), 181.

Notes: Total pig iron production of Alabama, Georgia, Tennessee, North Carolina and Virginia in 1850 was 54,405 tons or 9.65% of national output. Production in border slave states of Maryland and Kentucky was 67,886 tons or 12.04% of national output. U.S. total greater than 100% due to rounding.

wane prior to the Civil War; the smelting of recently discovered Lake Superior ores introduced new furnace concentrations in the Northwest with which Tennessee was poorly prepared to compete.[37]

Clearly, if the new company formed by Cravens and others was to succeed, it would have to be prepared to compete with the Western Highland Rim and the northern furnaces. However, production statistics illustrated that the smelting of iron in the Great Valley was on the wane in the 1850s. As a first move in its new venture, the company aligned itself with the railroads, a significant and accessible market for iron.

In February 1850 the East Tennessee Iron Manufacturing Company purchased a two-acre tract of land on the south side of Chattanooga.[38] Not coincidentally, the tract adjoined the main rail yard shared by the W&A and the N&C railroads. On this parcel the company erected a foundry and machine shop. There are no documents describing the facilities, but the foundry would have included a small cupola furnace for remelting pig iron for casting, and the machine shop would have housed various fabricating machines such as lathes, drill presses, and shears—all run by steam power.

By 1853, the company advertised that the foundry was

> prepared to execute orders of every description of cast, wrought Iron or Brass Work, at short notice and in the best manner. We are prepared to manufacture chilled railroad car wheels of the very best quality and freight cars of any description. Also, all other descriptions of cars or railroad columns, still and caps of any pattern desired for buildings. Also saw and grist mill castings of the latest and most improved kinds, Hotchkiss' water wheels, gin and crane gear, shafting pulleys and hangers, etc.[39]

The foundry and machine shop evidently found ready work, since they were equipped to build rolling stock (freight cars) from scratch.

Moving to Chattanooga in 1851, Cravens was in a position to personally supervise the company's newest venture, the construction of Bluff Furnace. The erection of a blast furnace at Chattanooga was a larger and much more involved operation than the erection of the foundry. Timberland in the vicinity of Chattanooga was needed for charcoaling operations. Iron ore could be barged downriver from the company's established ore banks near White's Creek, eighty miles by river from Chattanooga.[40] To do this, arrangements for the purchase or leasing of steamboats and barges had to be made. Last, but certainly not least, a reliable labor force, in-

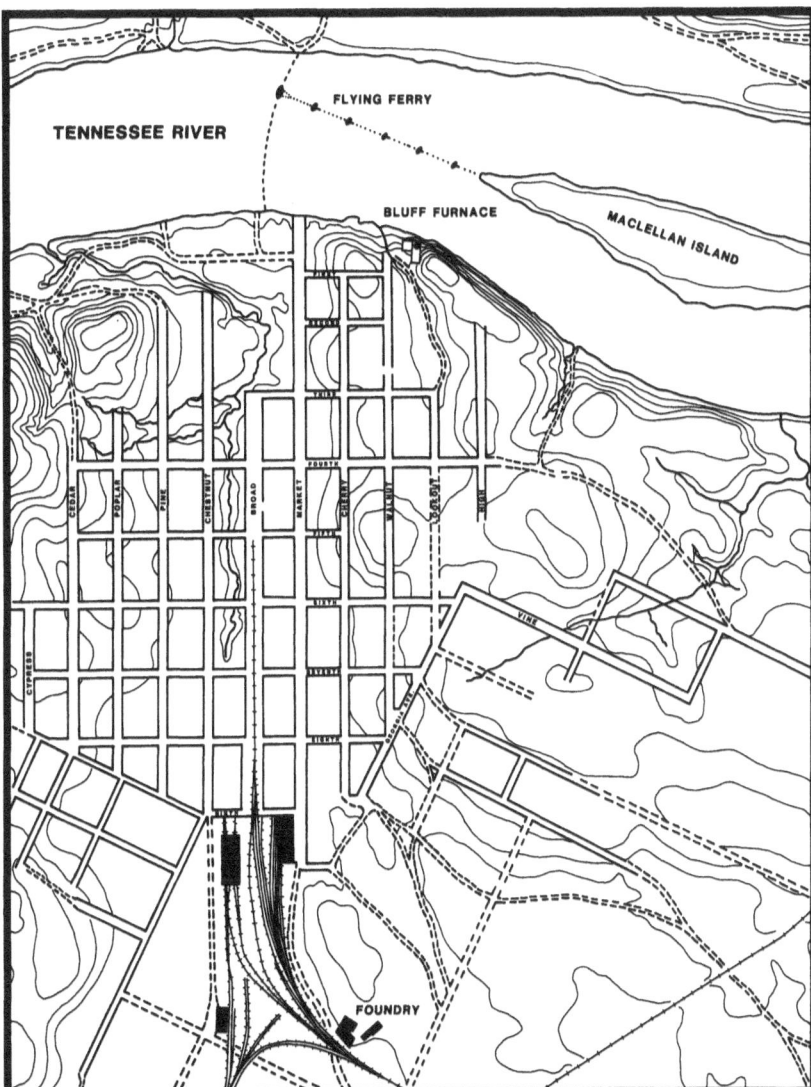

Fig. 3.3. A plat of antebellum Chattanooga. The principal facilities of the East Tennessee Iron Manufacturing Company were located on key transportation routes: the furnace on the river, and the foundry at the railyards. Adapted from the F.W. Dorr map, "Chattanooga and Its Approaches, Showing the Union and Rebel Works Before and During the Battles of 23rd, 24th and 25th November, 1863," from Dorr, *Chattanooga and Its Approaches*.

Fig. 3.4. Ross's Landing at Chattanooga. This engraving by artist Harry Fenn illustrates the character of the busy boat landing below the then-unchanged limestone bluffs that would later be sculpted to house Bluff Furnace. This sketch appeared in *Picturesque America*, vol.1:68, edited by William Cullen Bryant, in 1872.

cluding colliers, molders, and other skilled technicians, was needed to operate the facility.

On the south bank of the Tennessee River, just above the ferry landing on the north side of town, the site was chosen for Chattanooga's first blast furnace. At this point, high limestone bluffs commanding a view of midstream Chattanooga Island adjoined the sandy beach of Ross's Landing (fig. 3.3). On April 19, 1851, the company purchased a small triangular-shaped parcel bounded on the east by the bluff, on the north by the river, and to the south and west by the unimproved right-of-way of Lookout Street. The seller of the property was Jane Henderson, one of the town's earliest inhabitants.[41] More land adjoining this core parcel was needed to construct a full-sized plant. For these and other reasons, the East Tennessee Iron Manufacturing Company increased the limit of capitalization to one million dollars by an act of the Tennessee Assembly in February 1852.[42] Additional purchases in April and November of that year encompassed what would become the upper levels of the furnace site and its southern and western approaches.

The bluff property was conveniently located near a hub of the local transportation network (fig. 3.4). A swing ferry, using the natural current of the river to propel a ferry flat across the stream, was anchored at the tip of Chattanooga Island opposite the river bluffs.[43] The ferry landed just downstream from the furnace lot. From that busy beach, steamboats and flatboats were loaded and unloaded daily, serviced by horse-drawn wagons coaxed down the riverbank from the heart of town half a mile to the south.

The furnace lot abutted the west face of the high limestone bluff overlooking Chattanooga Island. This natural relief was ideal for the layout of the furnace plant. The furnace and casting shed would be built on a terraced slope some fifty feet above the normal river level to guard against the hazards of all-too-frequent floods. The bluff face behind the furnace stack was trimmed sheer so that the furnace could be placed against a vertical wall. The charging deck was built halfway up the face of the bluff, using natural rock ledges to anchor the trestlework of the deck. As a source of water power, the Tennessee River was unsuitable; the plant would be powered by steam.

We have no accounts of the actual construction of the charcoal furnace, but, as in any industrial enterprise, it was a period of heavy capital outlay.[44] Ancillary to the furnace construction was more real-estate speculation. In July 1853, J. J. James bought ten city lots from the company for the hefty sum of twenty-five hundred dollars. However, on the same day, and for the same amount, the company bought from James twenty-five acres of property south of the town, on Chattanooga Creek. This land was a prime industrial site, being adjacent to the mainline tracks of the N&C Railroad. The company had acquired an important asset, and the transaction suggests that industrial-site land speculation was an important aspect of the corporate operations.

In 1854 the furnace at the bluff was complete.[45] For reasons unknown, the furnace was not put into blast until 1856, perhaps because arrangements for a trained labor force and a reliable charcoal fuel supply had not been completed. James A. Whiteside had sold to the company in 1852, for an unspecified consideration, a tract of land on South Chickamauga Creek known as the Dearing Mills property.[46] On this 320-acre tract Gilbert W. Dearing operated several grist and/or saw mills. In November 1853, Dearing paid the East Tennessee Iron Manufacturing Company two thousand dollars for a one-third interest in two parcels of the mill tract on South Chickamauga Creek, the tract that bore his name. The deed notes that the interest in the parcels "makes the settlement including the South Chickamauga mills at present in possession of said Dearing."[47] The "settlement" on South Chickamauga Creek may have been a company town, perhaps the site of a colliery for the production of furnace charcoal. It was distant from the populous urban core area of Chattanooga, which had its own appetite for wood to heat and light homes, generate steam for manufacturing, and furnish building lumber.

Bluff Furnace was finally put into operation in 1856, burning the traditional charcoal as fuel, but employing a more advanced steam-powered hot blast, as did its companion furnace in Roane County. J. Peter Lesley's 1859 *Iron Manufacturer's Guide to the Furnaces, Forges and Rolling Mills of the United States*, which presents data gathered on southern furnaces in the period 1855-57, is one of the first documents to describe some of the technical aspects of the furnaces and forges in Tennessee. The following entries are taken from that compendium:

> 281. Eagle Steam and Water Hot-Blast Charcoal Furnace, No. 1, owned by the East Tennessee Iron Manufacturing Company, R. Cravens agent, and situated on White's Creek, sixteen miles west of Kingston, two miles north of its mouth opposite Jackson Ferry and White's Creek shoals, thirty miles west of Loudon, twenty-two miles west of Sweetwater the nearest station. Sixty miles by steamboat from the Nashville railroad at

Chattanooga with Memphis connections, was built in 1839, is 8 feet across the bosh by 33 feet high, and made in 1854 about 450 tons of metal (one-fourth) foundry out of dyestone or red fossil (Clinton, Upper Silurian, No. V.) ore outcropping along the south side of the Tennessee River. Stone-coal caps the mountain two miles off. Added its steam power lately.[48]

284. Bluff Steam Hot-blast Charcoal Furnace, owned by the East Tennessee Iron Manufacturing Company, R. Cravens Agent, Chattanooga, stands in Chattanooga, on the Tennessee River, under the bluff, three-quarters of a mile north from the railway station and thirty-eight miles by railway northwest of Dalton; was built in 1854, 10 1/2 feet across the bosh by 40 feet high, but made nothing until 1856 in about thirteen weeks of which year were made about 172 tons of metal out of fossil dyestone ore from Jackson's bank sixty miles up the river, near the dividing line between Roane and Meigs counties, three miles south of Eagle Furnace. The bituminous coal of the Raccoon Mines, now leased and worked by the Etna Mining Company, can be brought to furnace by railway; it is excellent for coke, and some thoughts are entertained of turning the present furnace into a coke furnace.[49]

The entries contain some provocative bits of information. The Bluff Furnace was completed in 1854 but was not brought into production until 1856. If the iron ore intended for use in the Chattanooga furnace was sixty miles (actually eighty-one river miles) distant by barge, there could conceivably have been supply problems. The quoted production for Bluff Furnace of 172 tons in thirteen weeks averages out to just under 2 tons of iron cast per day, certainly not a spectacular output for the period. Unfortunately, we have no production figures for the period 1856–59, the time span during which the conventional charcoal-fired Bluff Furnace would have shown a profit or loss to the company.

Lesley's *Guide* enumerated seventy-one furnaces, seventy-five bloomary forges, and four rolling mills that had been in operation in Tennessee after 1790.[50] Twenty-nine of the furnaces were in East Tennessee, and forty-two were in Middle Tennessee. Of these, only eleven furnaces in East Tennessee and thirty-one in Middle Tennessee were actually in blast or capable of operation at the time of Lesley's enumeration.[51] Some general characteristics of the state's blast furnaces can be drawn from the

Fig. 3.5. Bluff Furnace, 1858, drawing by David Hunter Strother. This engraving shows in some detail the layout of most of the furnace plant, including the ramp or incline to the charging deck level, and, in the foreground, a barge loading ramp for pig iron. From *Harper's New Monthly Magazine* 17 (1858): 289.

Lesley data (see Appendix 1), and this helps to place the events at Bluff Furnace in Chattanooga in perspective.

All of the furnaces in Tennessee enumerated by Lesley burned charcoal as fuel in the survey period of 1855-57. The blast furnaces of East Tennessee were almost exclusively water powered. The only two steam-powered blast furnaces employed in the region were both owned by the East Tennessee Iron Manufacturing Company: Eagle Furnace in Roane County and Bluff Furnace in Hamilton County. Eagle and Bluff furnaces were also two of the four East Tennessee blast furnaces using the hot blast in smelting operations. Of the thirty-one Middle Tennessee furnaces, only one employed water power to generate the blast; all the rest were steam powered. Only five of the thirty-one furnaces on the Western Highland Rim were equipped for the hot blast. In terms of blast-furnace technology, the East Tennessee Iron Manufacturing Company was one of the most advanced firms in the state.

Illustrator and writer David Hunter Strother visited Chattanooga during a tour described in a series of illustrated articles that appeared in *Harper's New Monthly Magazine* during 1857–58 under the title "A Winter in the South."[52] Strother, who used the pseudonym "Porte Crayon," sketched a detailed rendering of Bluff Furnace from a vantage point in the river near the ferry landing.[53] This pen-and-ink illustration documents the layout of the furnace in its charcoal-fired configuration (fig. 3.5). While some of the plant is lost outside the margin of the rendering, the casting shed, furnace stack, blast stove, and charging deck are clearly pictured. Also visible are two timber ramps from the riverbank, one steep incline running to the furnace top (used to load ore and fuel into the stack), and a second ramp on the riverbank off the end of the casting shed.

Lesley's *Guide* mentioned the possible future use of coke at Bluff Furnace. Cravens's unsuccessful effort in 1848–49 to use coke at Eagle Furnace must have made him a little gun-shy at the second attempt. Nevertheless, at Chattanooga, the coal deposits at the base of the Cumberland Plateau were relatively close and made even more accessible by the presence of the N&C Railroad running through the gorge of the Tennessee River. If the East Tennessee Iron Manufacturing Company were to adopt new production methods, the nearby coal fields could be tapped for coke.

The Etna Mines on Raccoon Mountain

Not surprisingly, the principals in the East Tennessee Iron Manufacturing Company were also investors in local mineral lands. In the antebellum period, all of Tennessee's blast furnaces (except Bluff Furnace) used charcoal as fuel. The use of coal or coke in foundry and blacksmithing operations was more common, and coal was increasingly being used for steam generation in a number of industries. There was, then, a small but growing antebellum market for coal in the South and the likelihood that, if the pattern of northern furnaces was followed, in time coke would be used in southern blast furnaces.[54]

In the 1850s James A. Whiteside, Ker Boyce, and Robert Cravens assembled title to an eighteen-thousand-acre tract on Raccoon Mountain in nearby Marion County. Bituminous coal was present in quantity on Raccoon Mountain, part of the larger landform of the Cumberland Plateau. The N&C Railroad passed through a gap immediately south of the coal lands bought by Whiteside, Boyce, and Cravens. This property held considerable industrial potential, and a new corporation was created to exploit Raccoon Mountain coal.

The Etna Mining and Manufacturing Company, incorporated by the Tennessee General Assembly on March 2, 1854, was chartered "for the purpose of mining and vending stone coal and other minerals, smelting and manufacturing the same, and engaging in any other branch of manufacturing, which said company may see fit to engage in."[55] Robert Cravens, David Rankins, Erasmus Alley, Ker Boyce, and James A. Whiteside were appointed commissioners to open books of stock subscription.[56]

In October 1855, James P. Boyce, J. A. Whiteside, and Robert Cravens leased their Raccoon Mountain coal lands to the Etna Mining and Manufacturing Company for a period of fourteen years.[57] The detailed lease agreement contained provisions for construction of a mining railway by the lessors, use of clay deposits on the property for making fire and common brick, permission to utilize iron ore on the property, and leave to construct smelters (blast furnaces) and rolling mills if the lessors decided to smelt the iron ore on Raccoon Mountain. The basic lease consideration or rent was twenty-five cents per long ton of coal (a 2,240-pound measure).

The Etna Mining and Manufacturing Company commenced development of what would thereafter be called simply the Etna (or Aetna) mines. In 1857, the company granted to Robert Cravens the exclusive right of mining coal on a 450-acre tract of land on the east side of the summit of Raccoon Mountain and "of taking therefrom a supply for not exceeding two blast furnaces and one Rolling mill to be located at or in the vicinity of Chattanooga."[58] Significantly, the leased parcel was adjacent to the mainline tracks of the N&C Railroad, the means by which coal or coke would find its way back to Chattanooga and other markets. The agreement specifically mentioned the possibility that some of the coal removed from the Etna mines might be coked on site.

A December 1859 mortgage note mentions improvements at the mine site as including: eight mules used at the mine; fifteen hundred tons of coal stored near the mine entrance; five thousand bushels of coke at the coal yard; two twelve-ton coal railroad cars, the office, scales, and coal trestle at the Chattanooga coal yard; the office, fence, scales, and coal trestle in the yard at Nashville; the fence and scales at Knoxville; the scales at Augusta; and all notes on debts owed to the company. The trust deed or mortgage demonstrates that already the company was shipping its product to several major cities throughout the Southeast.[59] Coke made from Etna coal was said to be of good quality for foundry use.[60]

The Etna Mining and Manufacturing Company eventually failed to make its lease payments and forfeited its lease on the Raccoon Mountain tract. In January 1860, Whiteside, James P. Boyce, and Cravens conveyed a fifty-year lease with the same general provisions as before to Samuel J. Agnew of Chattanooga. Even higher production quotas and minimum lease charges were specified. The 450-acre reservation held in trust by Robert Cravens was to be honored by the new lessee.[61] In May 1860, Agnew assigned the Raccoon Mountain lease to the Raccoon Mountain Coal Company, a corporation chartered in the state of New York in 1848.[62] Northern capital had made its appearance.

In November 1860, the Raccoon Mountain Coal Company executed a mortgage on the lease to secure funds borrowed to expand the mining operation of the Etna mines. By this date, some expansion of the operation is evident; the company maintained coal yards in Nashville, Knoxville, Chattanooga, Huntsville, Atlanta, and Augusta—all points accessible by rail.[63]

Etna coal was available throughout the Southeast by the end of the antebellum period, and the interests of the Etna Mining and Manufacturing Company and the East Tennessee Iron Manufacturing Company were interconnected. The proximity of the railroad to the coal mining operations on Raccoon Mountain was historically important. The Etna mines were the first in Tennessee to ship their products directly by rail from the mine site.[64] Later in the century, spur lines would be laid into the coal fields of the Cumberland Plateau to facilitate shipment and marketing.

From Charcoal to Coke

Few documents shed any light on Bluff Furnace's charcoal-fired operations in the period 1856–58. Moses Wells, who worked for the company nine years beginning in the winter of 1852, received seventy-five dollars a month to pilot the company steamboat, named *Union*, and barges hauling iron ore from the Roane County ore banks. Wells made flatboat trips

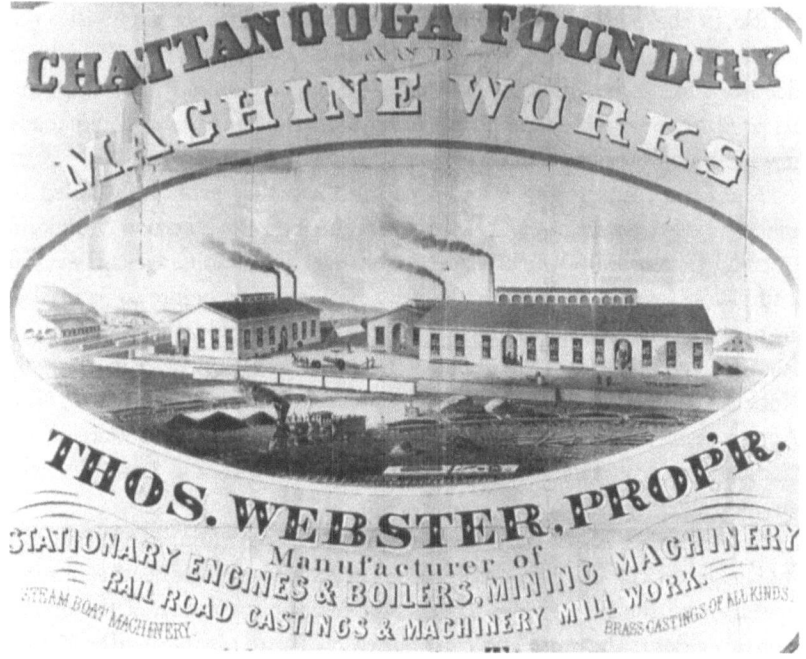

Fig. 3.6. The Chattanooga Foundry and Machine Shop. The earliest of the East Tennessee Iron Manufacturing Company's working assets in Chattanooga was this industrial fabrication facility. It was well sited to take advantage of the railroad trade; in the background at the left rear is the Union Car Shed, the passenger facility serving the N&C, W&A, and M&C railroads. This postwar view of Webster's foundry resembles the antebellum facility built by Cravens and company. From Lou L. Parham, *Parham's First Annual Directory of the City of Chattanooga*. Courtesy of the Chattanooga-Hamilton County Bicentennial Library, Local History and Genealogy Department.

every winter in order to haul pig iron downriver past the river obstructions at Muscle Shoals, Alabama, to make steamboat connections to markets in St. Louis.[65] What is not known is how much of the pig iron marketed in St. Louis was made at Bluff Furnace versus that made at Eagle Furnace.

As the 1850s drew to a close, several important changes in the operation of the furnace and foundry enterprises at Chattanooga occurred, some of them indicating that the company's local operations had lost momentum. In January 1856, the East Tennessee Iron Manufacturing Company leased the foundry operation off Market Street to the firm of Eastman, Lees and Company, an enterprise composed of partners John W. Eastman, Jonathan Lees, and Thomas Webster.[66] At two thousand dollars per annum, the lease included the machinery, smith shops, boiler shop, and cottage buildings on the foundry and machine shop lot (fig. 3.6). The *Chattanooga Gazette* advertised in early 1856:

> Chattanooga Foundry and Machine Shop—Eastman, Lees & Co.—
> The undersigned having leased for term of years, the Foundry and Machine Shops of the East Tennessee Iron Manufacturing Company, are prepared to execute all orders in their line at the shortest notice and on the most accommodating terms, viz: Stationary Steam Engines and Boilers, Mining and Mill Machinery of every description, Water and Gas Pipes, and Pump Castings, Car Wheels, Frogs, Switch Stands, and all kinds of Railroad Castings, Brass & c., Horse Powers, Threshing Machines, Corn and Cob Mills, Forgings of every description, Bridge and Car bolts. Particular Attention paid to the repairs of Locomotives and Cars. The Highest Price will be paid for old Copper and Brass.
>
> John W. Eastman
> Jonathan Lees
> Thomas Webster
> [of] Eastman, Lees & Co[67]

The five-year lease of the foundry and machine works prematurely terminated in May 1858 when the East Tennessee Iron Manufacturing Company issued bond for title to the property to Thomas Webster and R. D. Mann.[68] Webster and Mann, both English immigrants, evidently bought the lease from Eastman and Lees and secured sufficient capital to begin a series of yearly payments to buy the foundry in fee simple.[69] The reasons for this divestment of the foundry are not clear.

The year 1859 saw a major change in the configuration of Bluff Furnace and in its management structure. Two northern ironmasters arrived in Chattanooga and rebuilt the charcoal-fired furnace stack and ancillary equipment in order to use coke as a fuel. This conversion, which was successfully accomplished, marked the first such use of coke in a southern Appalachian furnace. The two men responsible for these regionally significant innovations were James Henderson and Giles Edwards.

Most of the evidence available indicates that Henderson and Edwards were part of a northern company that had leased Bluff Furnace from the East Tennessee Iron Manufacturing Company. No lease agreement has been found in the Hamilton County deed books, but such an agreement may not have been filed for record.[70] Many secondary accounts suggest that Henderson was personally leasing Bluff Furnace from the East Tennessee Iron Manufacturing Company;[71] others, that he was merely an employee at the managerial level. In an 1863 communication, Henderson indicated that the furnace "belonged" to him.[72] Correspondence cited below seems to suggest that ironmaster John Fritz of Pennsylvania was intimately involved with the Chattanooga operation, and other correspondence shows that Fritz was in communication in 1860 with iron mogul Abram Hewitt of New York about iron and coal properties in Tennessee. These scanty documentary details leave the exact composition and structure of the furnace lessees somewhat foggy.

A key question that eludes resolution is whether James Henderson converted Bluff Furnace to a coke-fired plant for his own (company) purposes or in response to a directive from the East Tennessee Iron Manufacturing Company. Recalling the manner in which Cravens had arranged for Edward Andreae to convert Eagle Furnace to hot-blast coke iron production, it may have been leased on an assumed-risk basis. In this scenario, the northern interests would have been, in effect, subcontractors of the East Tennessee company.

Of James Henderson as an ironmaster we know very little until after the Civil War. His name does not appear in the index to Lesley's 1859 *Guide*, an omission that shows only that he was not an owner, operator, or agent of a furnace, forge, or rolling mill at the time. The 1860 census lists Henderson's occupation as "proprietor," a designation that is somewhat ambiguous with respect to his status as a possible lessee. His postwar activities included obtaining several steel-making patents, suggesting he was more an inventor than a businessman or entrepreneur. However, in Chattanooga, it was clearly Henderson who was in overall charge of the plant.

Giles Edwards was born in England of Welsh parentage in 1824. Edwards immigrated to the United States and arrived at Carbondale, Pennsylvania, about 1842. He worked as a draftsman and pattern-making superintendent at Carbondale and Scranton, later supervising construction of a foundry at Tamaqua and superintending blast furnaces at Catasauqua. Edwards's health began to fail, in part due to overwork. Another rising star in the iron business, John Fritz, is said to have persuaded him to go south to recuperate his health.[73] His journey may actually have been to recuperate his fortune. Edwards wrote to Fritz in February 1858, complaining that he had been out of work for fifteen months, despite good references from Fritz, an individual widely respected in the Northeast.[74]

When Giles Edwards arrived in Chattanooga in the summer of 1859, James Henderson was already on the scene. From this period we have one piece of correspondence between Edwards and John Fritz, reproduced below with its original spelling and punctuation, discussing some aspects of the operation at Bluff Furnace:

Chattanooga July 18th 1859

John Fritz Esqr.
　　Dear Sir: I have deffered writing you, until the affairs of this concern had been definately settled.
　　Upon my arrival here I found the work which had been commenced on the furnace had been suspended by Mr. Henderson's orders, he being absent at the time, had left a note with Mr. Cravens for George, instructing him to prepare the estimate of the cost of repairing the old furnace, also a statement of the cost of moving it, or so much of the materials as would work in profitable in erection of a new furnace,—I concluded that it would be better to wait Mr. H's return, which ["took place" interlined] he did the next week Immediately on his return Mr. H wrote to N.Y. requesting Mr. Greacon to come down here at once, Mr. G got here in four days after the reciept of Mr. H's letter,—and after mature consideration it was concluded to let the furnace stand where it is and to make the best of it—it was agreed to that the Engine should be moved out on the bank directly at the end of the Boilers.
　　I find that it costs a great deal more to do work here than at the North— Webster asks 3 5/8 cents a pound for making all the castings his charges for machine work is proportionately high,—Mr. Greacon thinks that it is almost useless to think of asking favors of any man in the South & that it would be better to be prepared to be independent of all,—We will be under the necessity of making our own Firebrick, the clay at the Etna Coal Mines has been tested at the Loudon Rolling Mill and found to stand well.
　　Should Mr. C. Dorsey wish to come down here to line the furnace for us, we will make no other arrangements, when you see him next will you be good enough to mention this to him, and request him to advise me.
　　I have found the weather much warmer here than in Penn's the thermometer stands steady, at 95 degrees in the shade from 7 A.M. to 7 P.M, I do not know how the coke yard hands will stand it.
　　So much time has been lost, already that I am now kept very busy, please excuse my neglect in writing you,
　　　　I remain yours truly Giles Edwards.

The comments on the autonomy of the furnace strongly suggest that the smelting operation was being run entirely by out-of-state capitalists, supporting the view that the furnace was indeed leased to Henderson or a group of northerners. The taint of sectionalism is strong in the letter, hinting that the northern ironmasters were not entirely welcome in Chattanooga. The correspondence mentions two specialists in furnace construction—Mr. Greacon and Mr. Dorsey—being imported to accomplish construction of the new plant. Local expertise was thus lacking or, for business or political reasons, was not available to the northern ironmasters.

There were other portentous developments. The city council in 1859 granted the East Tennessee Iron Manufacturing Company the privilege of constructing a railroad line along Front Street from Lookout Street to Railroad Avenue and to use horse-drawn cars on that line in order to haul coal and other materials from the blast furnace west a quarter mile to the wharf area at the foot of what is today Broad Street.[75] Ideally, the wharf and the main rail yard at the south end of town would be linked by an industrial spur line, creating a rail and water transport linkage. The city fathers had, in fact, long since granted the W&A Railroad the use of the 126-foot-wide right-of-way of Railroad Avenue (aka Mulberry Street) in order for tracks to be run from the rail yards south of Ninth Street to the main wharf at Ross's Landing. The W&A never laid tracks closer to the waterfront than Fourth Street, however, in part because an enormous gully crossed the right-of-way.

Pig-iron production at the plant would necessarily be integrated with the barge transport of iron ore downriver from Roane County and the shipment of coke on the N&C Railroad east from the Etna Mines. The remodeled hot-blast coke-fired Bluff Furnace was to be part of a highly integrated enterprise operating at a scale far exceeding that of the traditional iron plantation so common in antebellum East Tennessee.

About 1860, a stereoscopic photograph of Bluff Furnace was taken from a point on the riverbank a short distance downriver from the plant. This important document (fig. 3.7) illustrates the furnace in its coke-fired configuration.[76] A cylindrical, iron-revetted cupola furnace replaced the massive stone charcoal stack. The furnace top was completely rebuilt, as was the charging deck. A brick hot-blast stove was set off to one side of the furnace and rested above the level of, if not in fact on top of, the casting-shed roof. A massive, tall smokestack, which does not appear in the 1858 woodcut of the furnace, is shown at the south end of the casting shed in the 1860 photograph; this stack evidently supplied a draft to the

Fig. 3.7. An 1860 stereoscopic view of Bluff Furnace. When compared to the *Harper's* view of the charcoal furnace, this photograph of the cupola-style stack and related equipment hints at the complexity of the conversions undertaken by Henderson and Edwards. Tucker and Perkins, Southern Stereoscopic Views, from the collection of Dr. James W. Livingood, Chattanooga Regional History Museum.

steam engines driving the blast or powering other equipment. Whether the stack was built in 1854 or in 1859 is not clear. Also absent in the 1860 photograph was the steep ramp from the river to the furnace top; this ore-hauling ramp may have been undergoing repairs at the time of the photograph, however.

The adoption of coke fuel at Bluff Furnace was a regionally significant development. In 1860, Bluff Furnace was using the most current technology to smelt iron. The blast was driven by a steam engine in contrast to older and less capitalized furnaces where unreliable water power was used. The pressurized blast was preheated before entering the blast pipes and tuyeres at the furnace base. Waste gas from the furnace top was vented to a specially constructed oven to heat the fresh blast air by convection. The fuel burned in the furnace stack was coke, a fuel with more thermal output pound for pound than charcoal and possessing a sturdy physical structure that permitted larger burdens in the stack. The cylindrical iron-revetted furnace stack was of the latest design. Replacing the massive stone charcoal stack in the form of a truncated pyramid was a rather more sparse, streamlined structure; surrounding the fire-brick core of the new coke furnace was a relatively thin sheet-metal cylinder. Bluff Furnace was the first in the southern Appalachian iron-producing region to successfully adopt coke smelting, and the 1860 photograph of the plant hints at its complexity.[77]

Joining the group of Yankee ironmasters in Chattanooga in 1860 was John Fritz, one of the principal figures associated with the early years of the Bethlehem Iron Company (which became the giant Bethlehem Steel in later years). Fritz wrote in his autobiography that he "went down to Chattanooga for the purpose of examining an iron ore and coal property" in the spring of 1860.[78] Undoubtedly, Fritz met with Edwards and Henderson in Chattanooga. One topic of conversation might have been the cupola-type blast-furnace stack, certainly the most novel of Bluff Furnace's new structural elements.

If Fritz did view the new Bluff Furnace, he might have done so with something more than passing curiosity. Upon his return to Pennsylvania, John Fritz left the Cambria Iron Works to superintend the construction of the Bethlehem Iron Company plant on the banks of the Lehigh River. Adopting the latest furnace-construction style, the first stack at Bethlehem was of the cupola variety: "The first furnace, or Number One, was built of plate iron one-fourth of an inch in thickness. It was the first shell furnace, as they were called, built in the Lehigh Valley. Iron was first made in this furnace on January 4, 1863."[79] Similarly, Furnace Number Two was of banded wrought-iron or crinoline construction.

John Fritz's prospecting visit to Chattanooga in 1860 would not be his last association with the iron industry of the town. The circumstances of his later connections would be very different, however. Another giant in the iron industry of the Northeast, iron mogul Abram S. Hewitt, had seen the report on the iron and coal property in Tennessee that Fritz viewed in the spring of 1860 and, in a letter to him in October of that year, concluded, "On the whole I would rather cast my lot North than South."[80] The passage of only a very little time would show the sad truth of that statement.

Unlike the principals in the story of Bluff Furnace at Chattanooga, the common furnace laborer is virtually invisible in the documents. The 1860 photograph of Bluff Furnace illustrates some of the furnace workmen standing at the furnace top, in the doorway to the casting shed, and others lounging on the adjacent slope as a wagon approaches the plant from the steamboat landing. Only in a few documents can we glean any information about these workmen and their lives.

The Eighth Census of the United States, based on population data collected during June 1860, gives us a glimpse of Chattanooga and its mechanical, foundry, and mining population sector. Listed in the "Free" population schedules of the census for Hamilton County are eighteen miners, one master machinist (Thomas Webster), nine machinists, two machinist apprentices, one mechanic, one nonspecific mechanical tradesman, four foundry molders, and three managerial-level persons: Robert Cravens, ironmaster; Giles S. Edwards, manager of furnace; and James Henderson, "Proprietor of Firm."[81] Of this group of thirty-nine individuals, only thirteen were born in the United States (see Appendix 2). The majority of Hamilton County's mining and manufacturing tradesmen were foreign born, most from the British Isles. Twelve were born in England, six in Wales, four in Ireland, and one in Scotland.

Hence, the majority of Chattanooga's skilled industrial work force was foreign immigrant labor, and those individuals who did not arrive in America with European iron-working skills probably came to Tennessee having learned their trade in the Northeast, most likely in Pennsylvania. The link between the iron industries of Tennessee and the Northeast was that of the diffusion of technology and innovations. One of the principal mechanisms of industrial diffusion was the movement of individuals. Some, like James Henderson and Giles Edwards, were more important than others to Bluff Furnace and the industrial development of Chattanooga. And in some cases, myth has overshadowed reality.

William Richard Jones, listed as a mechanic in Hamilton County in the 1860 census, later became a legendary ironmaster associated with an early and very successful Carnegie steel venture in Pennsylvania. Jones had worked at the Cambria Iron Works at Johnstown, Pennsylvania, and may have had a firsthand knowledge of coke-fired blast furnaces; the Cambria Iron Company had built four coke furnaces in 1853.[82] One biography specifically notes: "The following year [1860], as Master Mechanic, he went to Chattanooga, Tenn., to erect a blast furnace, but the menace of secession caused his return to Johnstown."[83]

Less authoritative accounts exaggerate the scope of his activities at Bluff Furnace. McGuffey's *Standard History of Chattanooga* notes, "At the outbreak of the [Civil] war, Capt. "Bill" Jones, representing Northern capital, operated a large blast furnace near the south end of the Tennessee River bridge. When news of hostilities reached Chattanooga, Captain Jones closed down the plant and went North."[84] Other accounts reveal a more human side of the Jones story.

Alabama iron historian Ethel Armes noted that Bill Jones had actually come south seeking the affections of one Harriet Lloyd, who was visiting the Edwards family in Chattanooga. He succeeded in surreptitiously marrying Harriet Lloyd, over the objections of Salinah Edwards, who had been charged to watch over her welfare.[85] Armes also states that Jones could not get a job at Bluff Furnace and instead opened a bar and billiard room.[86] If Armes is correct, his "northern capital" was invested in the "recreation" industry, not in iron.

Doubtless, the industrial work force of Hamilton County included many persons identified in the census only as laborers, individuals who worked at many jobs, perhaps on a seasonal basis. There is also another significant population sector to consider. One topic of considerable historical and anthropological interest is the employment of slaves in the furnace operation. Although the use of slave labor at Bluff Furnace has been intimated,[87] and may be taken for granted, no direct, primary historical documentation of their work at the plant has been found.[88]

The use of slaves in southern industry was an established practice, and one estimate puts the number of slave ironworkers in the antebellum South at ten thousand.[89] Slave labor had been successfully employed at the South's premier antebellum ironworks, the Tredegar Iron Works in Richmond, Virginia, but not without some resentment among the free white labor force and considerable reservation by the general populace of the city.[90] Free industrial workers felt threatened by competition with the slave labor force, but experience had shown that white managers and skilled labor could work alongside black laborers in the accomplishment of technical operations. The use of slave labor was a significant cost-saving factor in iron production at Tredegar.

Any concentration of slaves was a source of anxiety to white populations in the South. In 1856, rumors of a slave uprising on the Western Highland Rim of Tennessee led to swift suppression and violence.[91] The significance of this panic was that the supposed slave rebellion was confined to the large slave-labor work force of the furnaces and forges in the region. The slave insurrection panic of 1856 caused considerable disruption among the iron-making enterprises in Middle Tennessee, but the reaction, if any, in Chattanooga has not been assessed.

The Turning Point: November 1860

The year 1860 was a critical one in the history of Bluff Furnace and the East Tennessee Iron Manufacturing Company. Regional politics, the clash over the morality and economics of slavery, and the contest between the powers of the federal government versus the individual states were issues brought to a head in the fall of that year. The election of Abraham Lincoln galvanized regional politics, and the gears of war began grinding. The account of James Swank provides this commentary on the newly remodeled Bluff Furnace after its conversion to burn coke fuel:

> The new furnace was blown in in May, 1860, but owing to a short supply of coke the blast lasted only long enough to permit the production of about 500 tons of pig iron. All the machinery and appointments worked satisfactorily. The furnace was started on a second blast on the 6th of November, the day of the presidential election, but political complications and the demoralized state of the furnace workmen were obstacles too great to be overcome, and the furnace soon chilled from the cause last mentioned, and in December Mr. Henderson abandoned the enterprise and returned to New York.[92]

This account conveys two very important pieces of information: first, the initial coke blast stopped because of a shortage of fuel, not because of a technical or labor problem; second, labor unrest caused the termination of the final blast.

There is sketchy and contradictory data on the source of coke used at Bluff Furnace in 1860. It could be assumed that the nearby Raccoon Mountain coal mined by the Etna Mining and Manufacturing Company was being coked and shipped to Bluff Furnace, as previously cited documents have suggested. An 1897 mining report notes, however:

> Shortly before the war the Etna property changed hands, and the new management had a number of new ovens built to make coke for foundry use. This was the first effort of any size to make coke in the South. The

product was satisfactory for the uses intended; there was no thought of using it for making iron. Probably the first attempt to manufacture coke for use in blast furnaces in the South was by Robert Cravens. The coke was made in pits near where Rockwood now is, was boated down the river, and tried in the old Bluff furnace in Chattanooga.[93]

We know coke was on hand at the Etna Mines in December 1859, when the Etna Mining and Manufacturing Company used fifteen thousand bushels of the fuel to help secure a debt. But this supply of coke, by rail only fourteen miles away from Bluff Furnace, may not have been used at Bluff Furnace, perhaps due to its chemical composition. The Roane County coke was eighty miles away by river, a long journey by barge.

As for the second and final blast of the coke-fired Bluff Furnace, the account suggests that political unrest and labor problems brought the blast to a premature close. Friction may have existed between the northern manager, Henderson, and his southern work force, which very probably consisted of perhaps a dozen white technicians overseeing a larger work force of slaves. Whatever his role, his departure would have hampered a restart of the furnace; his technical expertise went with him. Giles Edwards's political leanings are unknown, but, either for political or practical considerations, he remained in the South during and after the Civil War.

The election of Lincoln might have balkanized sectional feelings and racial attitudes at the plant, resulting in a breakdown of organization and cooperation among the workmen and managers. And a furnace could be kept in blast only if all the machinery, materials, and tasks were religiously coordinated. Marcus B. Long, son of pioneer Chattanooga resident John P. Long, recalled in 1914 that Bluff Furnace "went out of blast in 1861 on account of the watchman going to sleep at his post and letting the iron chill and cool in the 'bosh' of the furnace, and put the furnace entirely out of business."[94] This description might contain several apparent errors, as with the date of the final blast, and positing the unlikely possibility that an active furnace would have been left in the charge of but a single watchman instead of a crew of fillers and topmen working under the supervision of the furnace manager or a founder. What rings true is the description of what happens when a blast furnace is not steadily fueled or the blast properly managed: the charge solidifies in the stack, rendering the furnace useless. As we shall see in chapter 4, the "chilling" of the furnace mentioned by Swank is an intriguing observation in light of what was discovered during archaeological explorations of the furnace site.

Difficulties at the Bluff Furnace plant paled in comparison to what would happen to the entire nation in the Civil War. At the outset of hostilities, many men rushed to enlist for what was thought would be a short and glorious fight. This phenomenon, called overmobilization, crippled vital industries in the Confederacy. To compete with the vastly superior numbers of Federal soldiers, the South drained the countryside of able-bodied men and left the industries of the region to fend as best they could. Although Giles Edwards, a practical ironmaster, and the older Robert Cravens both remained in Chattanooga, there is no conclusive data that the smelting of iron at Bluff Furnace resumed. Only one indication of a possible restart is found in the literature, in the account of this period by Armes: "No sooner had the Bluff Furnace been put into working order than in March, 1862, at the request of Judge Lapsely of Selma, whom he had met in New York, Mr. Edwards came into Alabama, and reconstructed the Shelby Iron Works."[95] The context of the statement does not indicate confusion of the first rebuilding of the furnace in 1859 with a possible second rebuilding in 1862. Thus, there remains the intriguing possibility that the furnace was rebuilt (if not put into blast) after the November 1860 chill.

The year 1861 is a void in the information on Bluff Furnace and the East Tennessee Iron Manufacturing Company. If Bluff Furnace was in working order in March 1862, when Giles Edwards departed for the Shelby Iron Works in Alabama, it apparently did not remain so for long. James Swank's commentary notes: "In the summer of 1862, before the Union troops took possession of Chattanooga, the machinery of the furnace was removed to Alabama by Giles Edwards, who used it in the equipment of a small charcoal furnace near the site of the present town of Anniston. This furnace was active for about two years."[96] Other testimony confirms movement of the blast equipment, but to a different locality. J. D. Harris, a miner resident in Hamilton County in 1860, later recalled of Bluff Furnace that "the furnace that we made the iron with was removed to Shelby, Ala., during the war."[97] Harris also recalled that the plant was owned by Henderson and Gleason of New York, but that Edward Giles was general manager and David Caldwin was furnace manager. The muddling of names after thirty years is not surprising; however, the names of Gleason and Caldwin appear in no other Bluff Furnace accounts, and their identities remain a mystery.[98]

But what equipment would have been removable from the plant? The most important elements of the furnace were the blast machine (the rotary fans or cylinder pumps) and the steam engine driving the blast machine. Piping and duct work for the hot-blast stove (a masonry structure that could not be transported) and downcomer pipe would have been reusable, as would the furnace tuyeres and related fittings. Tools and iron-handling equipment were also usable. This equipment would have been vitally important to the makers of Confederate iron.

There is little of record about the activities of the East Tennessee Iron Manufacturing Company in the year 1862. On March 29 of that year, the company conveyed an eleven-acre parcel of land on Chattanooga Creek to Lafayette Folger. The parcel, part of the tract bought from J. J. James in 1852, adjoined the tracks of the N&C Railroad and included one acre along the Tennessee River. On this prime industrial site Folger was supposed to build a rolling mill "for the purpose of rolling Rail Road and Merchantable Iron."[99] Later documents indicate that Folger was acting in partnership with Samuel B. Lowe, the individual whose name is associated with the actual construction of the rolling mill identified in the postwar period as the Vulcan Iron Works. Lowe, however, would not be allowed to complete construction of the mill. When Union troops occupied the city in September 1863, the incomplete plant was demolished as a war industry. Ironically, when the East Tennessee company bought the tract in 1852, it was probably with the intention of building a rolling mill on that site.

The stockholders of the East Tennessee Iron Manufacturing Company met in Augusta, Georgia, in April 1863 and agreed to a division and liquidation of the corporate assets. The number of stockholders, and the value of their assets, had dwindled since the formation of the company in 1847. Farish Carter had died in July 1861 and James A. Whiteside, president of the company, died in November 1861 in the process of retrieving an ailing son from Confederate military service.[100]

In April 1863, all of the company's assets were either in occupied territory or would be shortly, making it unlikely that the remaining stockholders would gain any benefit from them until the conclusion of the war. From the outcome of the meeting it is clear only that one stockholder had sufficient financial resources to buy out most of the company assets. The wealthy South Carolinian Ker Boyce had died a decade earlier, leaving an immense estate to the management of his son, the Reverend James Pettigru Boyce, who had acquired in his own right considerable interests in the company as well as in related mining enterprises.

The end result of the Augusta meeting was that the new president of the company, Robert Cravens, conveyed to J. P. Boyce, acting for himself and as executor of his father's estate, all of the property of the East Tennessee Iron Manufacturing Company excepting a parcel known as the Johnson Tract.[101] Bluff Furnace, Eagle Furnace and forge, the town lots in Chattanooga, the mineral and timber lands in Hamilton, Marion, and Roane counties—all went to J. P. Boyce, who obtained 5/13 interest for himself and 8/13 interest as executor of Ker Boyce. The value of the properties conveyed amounted to $65,750. There is no mention in the deed of James Henderson's owning the furnace or holding a lease on the facility.

The Civil War had come to lower East Tennessee, and one of its first victims was the East Tennessee Iron Manufacturing Company. The short life of Bluff Furnace at Chattanooga ended. During the first year of the war, Chattanooga had remained a quiet backwater. But military strategists on both sides realized the significance of the place to the territorial integrity of the Confederacy. Water and rail transport lines converged at Chattanooga, which sat adjacent to a major pass through the Cumberland Plateau. The fall of forts Henry and Donelson above Nashville early in 1862 signaled the invasion of the Confederacy. With its outer defenses gone, Nashville fell to the Federal army in February. Huntsville, Alabama, was occupied by Federal troops in April, severing Chattanooga's rail link with Memphis.

On June 7 and 8, a Federal detachment under General James S. Negley approached Chattanooga from the north and shelled the town from Stringer's Ridge on the north bank of the Tennessee. Although damage was minor, the vulnerability of the town was demonstrated.[102] A year later, forces under Colonel John T. Wilder, screening the southerly movement of Rosecrans's army moving to engage Confederate forces under Bragg, took up the same position on August 21, 1863, and bombarded the town for a second time. This incident predicated the evacuation of civilians and much of the town's industrial work force and machinery by rail.

Chattanooga was occupied without resistance by Union troops on September 9, 1863 (fig. 3.8). Later in the month, while pursuing the Confederate army under Braxton Bragg, Rosecrans's Federal army met with disaster at Chickamauga and fell back on Chattanooga. A curtain of fortifications was thrown up around the city, and the occupying army was besieged by the Confederates who held the surrounding high ground of Missionary Ridge and Lookout Mountain. The battles of the same names

Fig. 3.8. Union troops cross the Tennessee River from its north bank into Chattanooga, September 1863. Pursuing the Confederate army south, the Union armies passed over pontoon bridges into Chattanooga. Much of the industrial equipment in town, including the machinery of Bluff Furnace, had been evacuated south by rail prior to the occupation. At the furnace site, shown at the end of the pontoon bridge in this view, only building foundations are depicted. From Guernsey and Alden, *Harper's Pictorial History of the Civil War*, vol. 2:540.

late in November broke the siege and forced the Confederates to fall back slowly toward Atlanta. At the end of 1863, a Federal army stood in the open gateway to the heart of the Confederate States of America.

The year 1864 saw intensive construction activity by the occupying Federal army. Chattanooga was the staging area for the Federal assault on Atlanta; the town was transformed into a marshaling yard for thousands of troops and a shipping depot for enormous quantities of materiel. One of the construction needs of the Federal engineers was lime for mortar. Limestone rock, abundant in the region, was normally reduced to powdery form by calcination in masonry kilns. Of the ill-fated Bluff Furnace, James Swank's commentary tersely notes, "The stack of the furnace at Chattanooga was used by the Union troops as a lime kiln, by whom it was subsequently torn down."[103]

The furnace site was photographed during this period (about 1864), this time from the military bridge built just downstream (fig. 3.9).[104] Extensive demolition of the plant is evident in the photograph. To supplement the pontoon bridges across the river, the Federal army constructed a timber bridge across the stream, anchoring the midstream piers with tons of stone. It was reported that much of this stone was removed from Bluff Furnace; the massive cut-stone walls of the casting shed evidently disappeared into the bridge piers.[105]

The vicinity of the furnace assumed a martial air; on the crest of the bluff, to the east of the furnace site, a battery of cannons had been emplaced to defend the northern approaches to the town. In December 1863, James Henderson wrote to the commander-in-chief of the U.S. Army, Major General H. W. Halleck, to warn the Federal troops at Chattanooga of a potential threat to their fortifications on the bluff. Henderson relayed to Halleck that there was a cavern under the bluff that Confederate saboteurs might mine with gunpowder and explode, demolishing the bluff. Henderson's communication was written in Birmingham, England, where he was evidently studying British steel-making technology.[106]

It was in 1864 that another thread in the fabric of Chattanooga ironmaking was woven. John Fritz, then superintendent at the Bethlehem Iron Works in Pennsylvania, was contracted by the U.S. government to furnish machinery for a rolling mill under construction at Chattanooga by the United States Military Railroad. Fritz's brother, William, was appointed superintendent. The rolling mill, Chattanooga's first, rerolled rails worn out on the region's heavily trafficked railroads. Fritz's machinery performed the task at hand, and after the war the mill served Chattanooga's peacetime economy.[107]

Fig. 3.9. The "lime kiln" at the bluff, 1864. Demolition of the plant by Union troops left little stonework in place, as this photograph indicates, but the lower half of the iron-clad cupola-style furnace remained more-or-less intact—to be employed as a lime kiln. The delicate cast-iron arches supporting the sheet-iron stack encircled the tuyere and hearth levels of the furnace. The wooden shed at left was part of the lime-making operation. Detail from a photograph in the collection of the Jeffrey L. Brown Institute of Archaeology, University of Tennessee at Chattanooga.

The Etna mines on Raccoon Mountain had also suffered the indignities of war. The lessees of the property, the Raccoon Mountain Coal Company of New York, found themselves disenfranchised of their property at the outbreak of the conflict. The Confederate government had seized the mine and operated it until the Federal offensive in 1863. The advancing Federals evidently burned most of the miners' houses at the mines, unaware that the leaseholders were not Confederates but loyal Unionists. In October 1863, shortly after the capture of Chattanooga, the Raccoon Mountain Coal Company gave power of attorney to Milo Pratt of New York City and sent him south to confer with the military authorities at Chattanooga in an attempt to recover the mining operation.[108] In October 1864, the company transferred the mining lease to Pratt for a period of two years, during which time Pratt was to pay twenty-five cents per long ton of coal and conduct such repairs on the mining operation as he deemed expedient—unless hindered by rebel forces still in the field.[109]

Confederate Iron

The experiment at Bluff Furnace had not gone unnoticed in the troubled days prior to the start of hostilities. Production of pig iron with coke-fired hot-blast equipment had been successfully conducted at Chattanooga. Confederate iron producers, however, had much more on their minds than the introduction of new technology to their old plants.

When forced to rely upon its own internal industrial resources, and isolated from foreign sources by naval blockade, the Confederacy was faced with serious shortages that the progress of the war only made more acute. Many factors affected the performance of the heavy industrial sector of the Confederate economy, but one of the most acute was a shortage of pig iron. James A. Seddon, Confederate secretary of war, complained to President Jefferson Davis in January 1863:

> The most serious embarrassment to be apprehended in reference to the ordnance supplies is in the deficiency of iron. Before the war nearly all ironworks within the States of the Confederacy had languished or decayed, and from the sense of precariousness in the future and the scarcity of suitable labor it has been very difficult to establish them in sufficient numbers and on an adequate scale to meet the necessities of war.[110]

The Nitre and Mining Bureau was created on April 11, 1862, to coordinate the production of materiel for military manufactures, and in June 1863, all furnaces in the Confederacy were placed under its direct control in a desperate effort to meet the iron requirements of the several, and usually competing, branches of the Confederate military.[111]

At the outset of the war, the industrial potential of the Confederacy suffered serious setbacks. The slave states of Kentucky and Maryland had not been drawn into the southern union; these states produced sizable outputs of pig iron and foundry castings. The returns of the 1860 census revealed that Tennessee had ranked first in pig-iron production in the states that made up the Confederacy, with double the output of Virginia. However, much of Tennessee was in Federal hands after the first full year of war; the loss of pig-iron production from the furnaces on the Western Highland Rim was disastrous to the South's war industries.[112] Furnaces in Virginia were hard pressed to produce pig iron in any volume; the state was on the military frontier with the hostile Union and could scarcely defend its capital, much less outlying industrial plants.

The most important rebel industrial centers to emerge during the course of the war were the Tredegar Iron Works in Richmond, Virginia, and the Confederate works at Selma, Alabama. At the start of the conflict, the Tredegar Works and, to a lesser degree, the Etowah Iron Works in northern Georgia were the only plants in the South capable of producing large castings such as naval and siege cannon. Largely because of the state's isolation from invading Federal armies, iron production in Alabama was greatly increased and supplied a large proportion of the South's iron output for the duration of the war. During the period 1862-65, thirteen new furnaces were constructed in the state, and in all, seventeen furnaces turned their output toward the production of iron for the Confederacy.[113] Georgia's blast furnaces were of meager output, and the Federal advance on Atlanta in the spring of 1864 struck through the heart of the state's iron-producing region. Mark Anthony Cooper's premier iron manufacturing complex on the Etowah River vanished in smoke.[114]

One of the most important wartime iron-production centers in Alabama was the Shelby Iron Works in Shelby County. It was to this plant that Giles Edwards was called in March 1862 to repair the company's blast furnace. Edwards directed the construction of a second and much larger blast furnace, which came into operation in March 1863. The works also contained a rolling mill, which turned out rail and armor plate.[115] A Federal raid cut through Alabama's iron-producing region in the spring of 1865 and destroyed all but one of the state's blast furnaces. The raid reduced the arsenals, foundries, and military factories of the heavy industrial complex at Selma to rubble. Shortly after the raid, the Civil War ended.

The production of coke pig iron at Bluff Furnace had demonstrated that quality iron could be made with coke from southern bituminous coals. Coke pig-iron production was still considered a possibility, even during the war. Advertisements like the following appeared in regional papers such as the *Chattanooga Rebel*:

Blast Furnacemen Wanted

Who can use coke in making Pig Metal on a large scale. Wages—two hundred and fifty dollars per month. Advise immediately. F. A. Woodson, Selma, Alabama, or Capt. Wells, Chattanooga.[116]

In 1863, at the Cahawba Iron Works (Irondale) in Jefferson County, Alabama, a high-quality pig iron was produced using coke. This was the first coke pig ever made in Alabama, but shortages in the coal supply did not permit continuous production with that fuel.[117]

Reconstruction and Beyond

Bluff Furnace and the East Tennessee Iron Manufacturing Company had been consumed in the crucible of the Civil War. The furnace had been demolished, the company assets sold and dispersed, and the principals were scattered or dead. After the conflict, the surviving members of the Bluff Furnace enterprise resumed their individual pursuits.

James Henderson returned to the South in 1887 to attend a meeting of the International Association of Metallurgists and Mineralogists at Birmingham, Alabama. Henderson held several open-hearth steel patents of interest to local iron producers, who were convinced that steel production was the key to the further prosperity of the region.[118] The experimental Henderson Steel and Manufacturing Company was formed to test the notion that steel could be made from southern ores and coke. On March 8, 1888, the first basic open-hearth steel made in Alabama was cast.[119] This successful experiment introduced steel production into the Birmingham district and opened the economic floodgates of the region.

William R. Jones had gone north with his bride and, at the start of the Civil War, enlisted as a private. At the conclusion of the struggle, Captain Bill Jones returned to the Cambria Iron Works in Johnstown, Pennsylvania, and later became superintendent of the famous J. Edgar Thomson Works at Braddock, Pennsylvania. Jones rolled Pittsburgh's first steel rails in August 1875 and drove his furnaces to record-breaking levels of production. A legendary figure in the emergence of the Carnegie steel empire, Jones died an ironmaster's death in September 1889, the victim of a furnace explosion.[120]

Of the former Chattanooga ironmasters remaining in the South after the Civil War, perhaps Giles Edwards was the most successful. In 1862, Edwards and family had moved to Shelby County, Alabama. Edwards had rebuilt the furnace of the Shelby Iron Works. The quality of his work is remarked upon by Robert H. McKenzie:

> The Shelby furnace was technologically more advanced than most of its wartime contemporaries. The stack was built during the war by Giles Edwards, a native of Wales with long experience in the construction of

furnaces in Pennsylvania and Tennessee. The furnace was thirty-eight feet tall and was capable of producing twenty tons of iron per day. Consequently, its cost of operation was probably the lowest of any of the wartime furnaces.[121]

In 1866, Edwards again rebuilt the furnace, this time repairing the ravages of war. He participated in the construction of the Brierfield Iron Works in Bibb County, also in 1866, and subsequently acted as a purchasing agent and mineral prospector for the Pioneer Mining and Manufacturing Company. During this prospecting period, the Edwards family lived in a mansion at the Tannehill blast furnace.[122]

In the late 1870s, the iron industry in Alabama was booming. Edwards had begun construction of a charcoal blast furnace in Woodstock, Bibb County, in 1873, but the project was suspended until 1879, when the Edwards Iron Company was organized. The stack was enlarged to burn coke and was projected in 1880 to have an annual capacity of eleven thousand net tons of pig iron destined for mill use.[123] By 1883, a railroad had been constructed to Woodstock, and the plant was in full operation. Giles Edwards died during the economic depression of 1893, having spent the latter half of his life in service to the southern iron industry.

From his home halfway up the face of Lookout Mountain, Robert Cravens could look down into the valley where the town of Chattanooga stood on a broad bend in the Tennessee River. The level terrace on which Cravens had built his home had been the scene of fierce fighting during the Battle of Lookout Mountain on November 22, 1863. All that remained of *Alta Vista* was the timber framework of the house and a stone outbuilding. Following the capture of the town and occupation by Union troops, the Bohemian Club, a group of northern newspaper reporters, camped on the south brow of the mountain and watched as soldiers scavenged through the remains of the house.[124] What had not been broken by shot and shell was taken as souvenirs. As the tide of war shifted south toward Atlanta, the reporters abandoned what they called "Camp Harper's Weekly," and the Cravens family returned to the ruins of their home.[125] Around them, and beneath them in the Chattanooga Valley, were the remnants of a place transformed by war.

For ironmaster Robert Cravens, a decade of work in lower East Tennessee had vanished. Cravens was by no means destitute at the end of the war, owning as he did considerable acreage in Hamilton and Marion counties. Nor was he of the type to be crushed by adversity. After the war, he invested in government surplus wagons which were auctioned off in Chattanooga by the thousands.[126] Profits from the resale of the wagons and, presumably, income from his real estate (including Etna Mines) put him back on his feet financially. He was president of the Chattanooga Leather Manufacturing Company in the 1870s[127] and later organized the Southern Manufacturing Company, of which he was president. After a lingering ailment, Cravens died on December 3, 1888. His funeral was delayed several days by a snowstorm that blanketed the valley in white, but he was finally laid to rest in Forest Hills Cemetery in the shadow of Lookout Mountain.[128] The family home, *Alta Vista*, was eventually restored and is now a national historic landmark.

The Archaeological Transformation

Bluff Furnace, when put in blast in May 1860, was the first of the lower South's modern pig-iron plants. Its coke fuel, iron shell furnace, and steam-powered hot blast represented the pattern in blast-furnace construction and practice that would, with a few technological improvements, persist well into the twentieth century. Historical circumstances prematurely terminated the operations of the plant long before its viability could be tested.

By the end of the Civil War, Bluff Furnace, like most of its contemporaries in the war zone, was in ruins. Only the base of the furnace survived; all else had been demolished. And this too, in time, disappeared, possibly broken up for scrap and remelted in some local foundry. One of the last references to the furnace came in March 1867, when the Tennessee River crested in the highest flood in the recorded history of the valley; flood waters touched the "lime kiln" on its terrace fifty feet above the normal pool.[129] During the flood the nearby military bridge was swept away, and the piers containing the stonework from Bluff Furnace were thrown down by the waters.

In the decades after the war, memory of Bluff Furnace and the East Tennessee Iron Manufacturing Company slipped steadily toward oblivion. In July 1886, the East Tennessee Iron Manufacturing Company issued a revised deed correcting some errors in the 1863 division of the corporate assets. At that date, Robert Cravens was still signing as company president.[130] The company quit-claimed title to the Deering Mill place on South Chickamauga Creek to the trustees of Ker Boyce's estate in October 1889. In this latter

Fig. 3.10. Bluff View in 1905, from a tinted photographic postcard published by the Rotograph Company, New York. A real estate boom in the late 1880s saw the construction of stately mansions on the bluff top, with more modest houses being erected on the fringes of the limestone cliffs. The story-and-a-half bungalow at lower right rested on the charging deck level of the Bluff Furnace site. At this date, the furnace and casting-shed areas of the plant were little more than a ravine.

transaction, Tomlinson Fort signed as president of the company, Cravens having died.[131] This is the last recorded transaction of the company uncovered to date. After 1863, the company existed only on paper.

Chattanooga enjoyed a minor real-estate boom in the last years of the 1880s, and Bluff View emerged as a prime residential area, having a panorama of the river. The waterfront adjacent to the bluff was also valuable industrial property, given its proximity to the steamboat landing at the foot of Market and Broad Streets. The site of Bluff Furnace was subdivided into residential and commercial parcels. In the late nineteenth century the furnace lot lost its historical industrial identity.

Construction of the Walnut Street Bridge across the Tennessee River downstream of the bluff was completed in 1891 and served to increase the isolation of the furnace lot from the surrounding neighborhood. One of the principal bridge piers was constructed in the right-of-way of Lookout Street just west of the former casting shed. Between 1889 and 1904, a small house was built on the former charging-deck level of the furnace site. Retaining walls had been built to terrace and stabilize the west side of the bluff. This residential development is shown in a hand-tinted postcard of 1905 (fig. 3.10). A river gauge was mounted on the river face of the bluff, and fences were situated to protect children from the precipitous slopes.

The residential character of Bluff View remained intact for most of the twentieth century. In 1970, houses adjacent to the furnace site—including the one on the charging-deck level of the furnace—were condemned and demolished as Riverside Drive was extended along the Chattanooga riverfront. Spoil dirt from the grading of the right-of-way was pushed downslope over the furnace site. Soon the area was covered in dense vegetation and frequented only by anglers on their way to the shoreline. After a few years, a storm sewer outfall had cut a deep erosional gully in the fill embankment and exposed the massive limestone doorway to the furnace casting shed. This was the beginning of the second life of Bluff Furnace at Chattanooga.

CHAPTER 4

The Archaeology of Bluff Furnace

The discovery of Bluff Furnace in 1977 was both accidental and purposeful. By chance, water running downhill from a storm sewer had cut through several feet of soil and exposed some wall foundations. Only when combined with Jeff Brown's background knowledge and research on the site did this chance encounter become significant. Besides identifying the foundations as belonging to Bluff Furnace, Brown's initial excavations indicated that furnace remains existed under a prodigious amount of fill. However, the extent and condition of the archaeological remains were unknown. Brown's research had located important pictorial representations of the furnace during, and just after, its period of operation; combined with the archaeology he had completed, these data meant we had some idea of what to expect before beginning to dig.

We became involved in the project just after Brown's death. As the newly installed director of the Jeffrey L. Brown Institute of Archaeology, Nicholas Honerkamp was expected to "continue where Jeff left off"; in essence, he had inherited the Bluff Furnace research program. Having worked with Brown on earlier projects, although not on Bluff Furnace, M. Elizabeth Will and R. Bruce Council[1] were aware of the site's promise as well as its problems. The first order of business was to prepare the site for large-scale excavation. To this end, Honerkamp organized a series of what were euphemistically called "volunteer sessions" on weekends during the winter of 1981 to clean the property of accumulated debris and to cut down and remove the numerous large trees that physically prevented excavation.

A necessary psychological characteristic of any archaeologist is the unshakable conviction in one's ability to do just about anything—no matter what it is—that contributes to the completion of an excavation. Perhaps this is what led Honerkamp to believe that he could safely remove large trees on a twenty-degree slope without the assistance of professionals. It was undesirable, and probably illegal, to simply fell the timber and allow it to roll into the Tennessee River, so elaborate cutting procedures were employed to induce the trees to fall uphill, where they could be chain-sawed into manageable sections and removed. A stubbornly uncooperative oak that was leaning in the wrong direction nearly brought a premature end to both the research and the research director during a particularly notable work session. A heavy rope had been attached to the top of this tree in an attempt to influence the direction of its fall. After the rope was tightened and tied off and all was in readiness, the oak was cut at its base—but it didn't come crashing down. Instead, the tree remained upright by spinning around about ninety degrees on its base, in a manner reminiscent of a climactic scene in Ken Kesey's *Sometimes A Great Notion*.[2] As Honerkamp began to give the crew directions on what to do next, a branch that had been forced back under tremendous pressure by the spinning motion of the trunk suddenly let go, striking him in the face in mid-sentence. He required over thirty stitches and still retains an impressive "dueling scar" on his right cheek to remind him of industrial archaeology and human frailty.[3] This incident soon led to the hiring of a professional tree-removal firm, and the largest trees were all safely cut down.

Solving one problem immediately created another. Large trees lying on the ground are just as much a hindrance to an archaeologist as upright trees. The treacherous terrain, combined with the large diameters of the trunk sections (some in excess of three and one-half feet), made removal attempts seem like exercises in Russian roulette. It was at this point that several private construction firms came to the rescue by donating heavy equipment for use at the site. The largest tree sections were dragged out by using steel cables attached to a fifteen-ton hydraulic crane (complete with a skilled operator), and a dump truck transported them to the nearest landfill. Along with a twenty-ton track hoe, this equipment was also useful in removing old washing machines, car parts, stoves, and other items that littered the property.

From Brown's earlier work it was obvious that hand excavation of the furnace remains would have to be preceded by removal of the extensive modern overburden, preferably with heavy machinery. The initial attempt was made with another corporate donation, the uniquely appropriate backhoe shown in figure 4.1. Known as a Menzi-Muck Climbing Hoe, this highly specialized machine was developed for use on the sides of mountains and in shallow rivers. It is relatively light but powerful,

Fig 4.1. The Menzi-Muck Climbing Hoe. On slopes too steep for conventional backhoes, this specialized piece of machinery was used for initial removal of modern overburden and to create working terraces for other heavy equipment. The use of such machinery on industrial sites is typical; the scale of the digging tools is in proportion to the nature of the site.

with telescoping front legs instead of wheels to enable it to remain level on a severely sloping surface; the back wheels are likewise vertically adjustable. The bucket is attached to an arm that also has telescoping capability. While it is slow moving because it lacks motive power (it must be dragged to its destination by manipulating its own bucket and arm), it proved to be extremely useful in the initial preparation of the site, especially for cutting access ramps for other, more traditional types of earthmoving equipment.

Of course, the goal of all this intense effort and exotic machinery was to make archaeological fieldwork possible. The site was finally excavated during the summer and fall of 1981, first under Honerkamp's supervision with University of Tennessee–Chattanooga students enrolled in a seven-week archaeological field-school, then for six weeks with a professional crew under the direction of Council. In 1983 Honerkamp again returned to the site with a UTC field-school for two weeks of limited excavations. The results of this research were previously presented by us in a 1982 technical report as well as in a 1987 article by Honerkamp.[4]

Research Design

Archaeologists usually begin an excavation with a set of questions to be answered and an excavation plan designed to generate the answers. These questions and approaches are what constitute an archaeological *research design*, while a project's *research goals* are defined and then (one hopes) met by applying a particular *research methodology*. At the nitty-gritty level, *field techniques* are the actual means by which dirt is moved, artifacts collected, and features recorded.

Any number of methodologies and approaches can be used at a particular archaeological site. The choice of one over the other derives from a combination of factors—including the research goals established for a particular project; the available time, money, and effort that can be directed toward meeting those goals; and the site's physical condition. At Bluff Furnace the main goals were to determine the extent and condition of the physical remains associated with the antebellum furnace complex. Despite Brown's promising results, we were not entirely sure that substantial intact and interpretable archaeological remains were down there. Enough had happened at or near the site over the previous century to raise serious doubts about the success of any research program to be carried out. An erosional ravine had already cut through and destroyed some of the buried remains; the furnace was documented as having been torn down prior to 1863 (recall the discouraging photograph of the site taken during the Union occupation of Chattanooga); a turn-of-the-century home had been built directly over part of the site; the construction of the Walnut Street Bridge in 1891 might have resulted in the recycling of the site's architectural remains (foundation stones could have been used in the bridge's

Fig. 4.2. The doorway in the north casting-shed wall. Coauthor M. Elizabeth Will examines the massive stone footings and doorway in the casting shed foundations. The lower portion of the wall was undressed stone while the finished upper portion, originally exposed aboveground, consisted of finely dressed squared blocks or ashlar masonry. The doorway was ten feet wide.

footings); and Riverside Drive, with its attendant utility and sewer lines, was built just south of Bluff Furnace in the late 1960s.

In order to realize the overall goals, the excavation strategy was aimed at uncovering as much of the site as was feasible, given its unique topography. This required the removal of a tremendous amount of modern fill, a task that could never have been accomplished without the donation and use of several pieces of heavy earth-moving equipment. Decisions about where to locate particular test units were based on a model of the site's presumed layout that had been constructed from the combination of period photographs, contemporary descriptions, and the earlier test excavations. As more and more of the site was exposed, this synthetic model became increasingly detailed and accurate. Eventually a positive feedback loop of combined archaeological and documentary data was created, allowing highly accurate predictions to be made about where major site features were located.

After we had established the existence of archaeological remains, the research goals could then be fine-tuned. The next objective was to locate several presumed *activity areas* where distinct behavior relating to the operation of the furnace took place. One of the advantages of working with industrial sites is that it is sometimes possible to establish an unambiguous, one-to-one link between the archaeological remains and the human behavior that generated those remains. Except under special circumstances (Pompeii comes to mind), this usually is not the case at nonindustrial sites. The combined documentary, photographic, and archaeological information indicated that the following activity areas were present at Bluff Furnace.

1. A *charging deck* or working platform from which the furnace stack was filled with ore, fuel, and flux. Clearly indicated in the 1858 woodcut and 1860 photograph were a roof over the deck and several auxiliary buildings at the top of the limestone bluff. Evidence of these structures as well as examples of the furnace stock might be found here.

2. The *steam-boiler area*, approximately fifty feet south of the charging deck and at about the same elevation, where a steam boiler and associated smokestack were thought to be located.

3. The *casting shed*, where the molten iron from the furnace hearth was routed into sand beds and cast into sows and pigs. Besides indirect documentary data, Brown's earlier excavations and the extant foundations exposed in the erosional ravine indicated that substantial limestone foundations of the casting shed walls were present south of the furnace compound, as shown in figure 4.2. Casting by-products, such as sprue fragments and splash iron, were also expected to be recovered in this area, as was direct evidence of the casting beds.

4. The *furnace base*, the critical functional area and the "heart" of the site, located adjacent to the foot of the bluff. Since two distinct types of furnaces were known to exist at Bluff Furnace, we hoped that traces of both would be present; comparison of the woodcut and photographs of the site led us to believe that the remains of the two furnaces were probably in close if not direct proximity to each other. By-products of the furnace operation, such as charcoal, cinder, and slag might also be expected near the furnace base.

5. A *slag pile* adjacent to the furnace. Small quantities of slag had been noted by Brown next to the casting shed. The most logical place to dispose of this unwanted material would be downslope from the furnace complex.

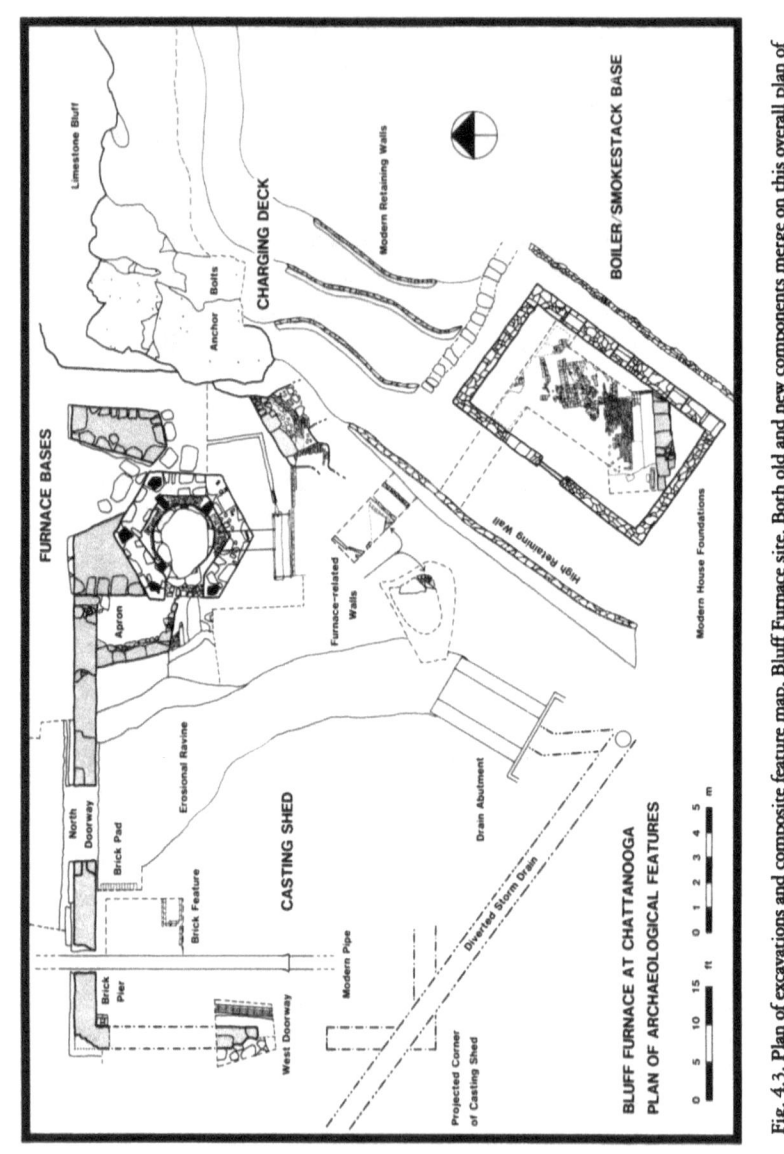

Fig. 4.3. Plan of excavations and composite feature map, Bluff Furnace site. Both old and new components merge on this overall plan of the furnace site. At lower right are rectangular foundations of the turn-of-the-century cottage built on the charging-deck level of the furnace. Within its foundation walls are remains of the steam engine mounts. A series of retaining walls of late nineteenth- and early twentieth-century origin are staggered across the bluff slope. Similarly, modern storm sewer pipes and outfalls cover the western portion of the site. The casting-shed walls and doorways are at upper left, and the bases of the charcoal and coke furnaces are top center.

The horizontal locations and relationships of these areas are illustrated in the figure 4.3 composite site map.

Methodology

All of the areas listed above are of a fairly substantial size. With the exception of the slag pile, excavations of only a small sample would probably not suffice to give a clear understanding of the form and function of each activity area. Hence, the decision to undertake large-scale excavation was not a self-indulgent luxury but rather a research necessity if an accurate understanding of the site's distinct components was to be achieved. Given the large amount of recent fill over the archaeological remains, the use of carefully controlled earth-moving machinery likewise was a practical necessity. At all times the trackhoe, front-end loader, and backhoes were run by skilled operators under the ever-watchful eyes of the UTC archaeologists. After the removal of the overburden was completed, more traditional hand excavation was carried out, as seen in figure 4.4.

Since excavation, no matter how carefully done, is by nature a destructive procedure, it is essential that the data resulting from this destruction be carefully recorded and preserved. Without the documentation, the artifacts recovered are virtually worthless for research purposes. More than anything else, it is this critical step in the excavation process that separates the archaeologist from the casual "artifact hound," the not-so-innocent bottle collector, and all the other myriad forms of looters who, with metal detectors and shovels in hand, prey upon archaeological sites. Field documentation at Bluff Furnace included photography, notes, mapping and recording of the locations and contexts of artifacts collected for analysis. Seventy-two rolls of black-and-white and color photographs and thirty-eight *plan view* and *profile* maps visually reproduced the features and stratigraphy discovered at the site.[5] A field-specimen catalog was kept to maintain provenience information (three-dimensional location data) on the artifact collections. *Features*, defined as any natural or cultural disturbances to the ground, or immovable artifacts such as foundations, etc., were numbered in sequence within each activity area. Data on their size, location, composition, and probable function were recorded. To ensure precision in vertical measurements, a transit and stadia rod were used. With over forty feet of vertical relief to contend with at the site, it was necessary to establish six semipermanent transit stations at different lev-

Fig. 4.4. Hand excavation in the furnace area followed the removal by machinery of the modern overburden. At lower left is the coke furnace base. Working in the shadow of the high limestone bluff, crew members were required to wear hardhats to protect them from falling debris.

els during the course of the excavation. Tied in to a permanent benchmark at the top of the bluff, elevations for all features could be given in absolute elevation above mean sea level. Extensive narrative notes were kept by the supervisors, who recorded excavation procedures, observations, measurements, and other essential data.

Archaeologists not only discover and record features during an excavation—they also collect artifacts. The kinds of artifacts that are collected and the way they are recovered depend on what research goals are defined, coupled with practical considerations. At Bluff Furnace a 100 percent collection policy for most iron artifacts was adopted. However, we decided against screening the fill dirt, even though this probably would have increased the recovery rate of small items. The subsoil and redeposited fill at the site consisted primarily of orange clay and limestone rocks, and if screening had been attempted, we would still be in the field. Due to the volume, it was not feasible to collect all fuel, ore, and slag fragments. Instead, limited samples were taken from well-defined features and contexts for chemical analysis. All complete firebricks, common bricks, splash iron, and dressed limestone blocks were retained for future reuse in the planned reconstruction and exhibition of portions of the site. Because we could not think of any particular exhibition or research value that brick fragments might possess, we decided not to collect them at all.

The following descriptions of the recovered artifacts and features are divided according to the activity-area model described above. Starting at the charging deck at the uppermost level of the site, we will work our way down, literally and figuratively, to the furnace foundation and casting shed. Although the site was excavated using metric measurements, as is standard practice in American archaeology, only the English equivalents appear here in the interest of readability. Another argument in favor of the English system is that it was used by the site's original inhabitants; the reported dimensions and distances are more intelligible under a nonmetric system. By contrast, the sizes of some of the excavation units may seem odd, but that is because they were originally dug in metric dimensions.

The Charging Deck

Located on top of the thirty- to thirty-five-foot-high limestone bluff adjacent to the furnace base, the excavated portion of the charging deck measured about seventy-five square feet. Actually, "excavation" is somewhat of a misnomer for this part of the site. Jeff Brown had tested this area in 1977, so the later efforts consisted mainly of clearing off brush and removing about a foot of duff and humus that had washed down from the slope above. The bluff itself is made up of several layers of limestone that slope down slightly to the southwest. Limestone solution pockets, channels, and terrace-like steps are present on its surface. The exposed faces of the bluff clearly exhibit numerous drilling scars, indicating that overhangs in these areas were cut away to create a more vertical configuration. The top of the bluff seems also to have been trimmed to create a terrace. Another modification was made to a large vertical crevice in the west bluff face. Near the crevice bottom was a large stone pad that served as the foundation for two other features (fig. 4.5). These were a curved brick retaining wall that closed the face of the crevice, and a remnant of

102 / Industry and Technology in Antebellum Tennessee

Fig. 4.5. Stone and brick construction below the charging deck. Situated southeast of the furnace base was a massive stone pier supporting a mortared brick feature probably built close to the face of a natural fissure in the face of the bluff. In the foreground is an iron reinforcing frame probably associated with the hot-blast stove.

another brick wall that began at the level of the furnace base. This latter feature is discussed later in the furnace sections.

The top of the bluff was carefully cleared by hand, first with shovels and then with brooms. Exposure of the bluff revealed remnants of twenty-nine iron bolts or rods set vertically in the limestone. Nine empty holes were also noted, but these could have resulted either from additional rods or from stone cutting efforts when the bluff was terraced. The rods were secured to the bluff by first drilling an oversized hole in the stone, placing the butt of the rod in it, and then filling the surrounding cavity with molten iron. When the iron cooled, the rod was firmly anchored in the stone. As can be seen in figure 4.6, the rods came in two sizes. The large

Fig. 4.6. Anchor bolts at the top of the bluff. Two sizes of wrought-iron anchors were set into holes drilled into the rock shelf directly east of and above the furnace base. The larger anchors were threaded for bolts, evidently securing the superstructure of the charging deck to the bluff top. In the upper right background is the north doorway of the casting shed. The photographic scale in this and subsequent photos, unless otherwise noted, is in ten-centimeter zones, for an overall length of 1.6 feet.

version was 1.4 inches in diameter and threaded on the end to receive nuts. Smaller rods of just under 1 inch diameter, many of which were broken off at the surface, lacked the threads. Several complete examples of both types were over a yard in length. Most were bent over toward the northwest, suggesting that their deformation took place all at one time, probably during a demolition event. The 1864 photograph (fig. 3.9) shows that the charging-deck structures were gone by the time the photograph was made.

In all likelihood, the iron rods we discovered were used to anchor the buildings and, in the case of the large bolts, some of the machinery to the surface of the charging deck. A heavy-duty winch would have been necessary to haul stock up the ramp illustrated in the 1858 woodcut (fig. 3.5) of the site, and the large anchor bolts found near the edge of the bluff were well suited to this purpose.

The Steam-Boiler Area

Lying south of the bluff top at approximately the same elevation, the steam-boiler area was excavated during the 1981 fieldwork season and again in 1983. A two-story residential structure, complete with basement, had been constructed in this location sometime between 1889 and 1904, and it was demolished prior to construction of the Riverfront Parkway in the late 1960s. To the northwest of the building was a substantial stone retaining wall, about ten feet high, that provided a level surface between the wall and the building. The slope around the house had also been terraced with a series of knee-high retaining walls. Both the house and the large retaining wall are readily apparent in the 1905 postcard reproduced in figure 3.10.

Initial excavation of the steam-boiler area was carried out during the 1981 field-school. Since we knew that the house structure was a late addition to the site and not associated with the industrial component (it was demolished in the 1960s), this was considered to be the most appropriate area for inexperienced students to start with: they could develop their knowledge and experience in archaeological fieldwork in a noncritical part of the site. A test pit was measured out over the apparent northwest corner of the rectangular structure, and the arduous task of removing demolition fill began. The top of the stone foundations was encountered beneath about a foot of fill, as were a brick runoff drain and a stone stairway on the building's north exterior, and copious amounts of early- to mid-twentieth-century domestic debris within the structure. Further excavation revealed the presence of a deep rubble-filled cellar, and the full extent of the twenty-by-thirty-two-foot rectangular foundation was traced out.

For safety reasons it was decided to relieve the upslope burden on the large retaining wall. Consequently, after the field-school was over, the institute crew used a trackhoe to completely remove and discard the demolition fill from within the structure. The earthen floor enclosed by the basement foundations was exposed, and grading around the house rec-

Fig. 4.7. Brick structures in the steam-boiler area. Several trenches were dug within the stone foundation walls of the turn-of-the-century cottage built on the charging-deck level of the furnace. Exposed are a furnace-period foundation to the right, and, at center, curb and dished floor elements of brick probably associated with ash basins under the firebox of the steam-engine boiler.

reated the surface contours of the residence when it was occupied. The foundation walls were about a foot and a half thick and formed of mortared limestone blocks in random coursing. The highest surviving point of the wall was well below the original first-floor level of the house. Three small window sills were noted in the southeast wall, and a large doorway—the same one shown in the 1905 postcard—opened through the center of the northwest wall.

Although excavation and clearing of a modern house foundation did not contribute directly to our knowledge of the antebellum furnace, this effort proved to be a means to a very valuable end. Discovered in the surface of the earthen floor of the basement was a large stone foundation. This feature ran at an almost perfect diagonal across the southwest corner of the basement, and it had been slightly *truncated* (or cut into) by the construction of the later house foundations.[6] The orientation, size, and construction materials used in the earlier, lower foundation were all solid evidence for a furnace association: the foundation was oriented to the cardinal directions, as were the casting-shed walls and furnace; the two-foot, three-inch-thick wall was the same width as those of the casting shed; and the mortar used was of a distinctive type identical to samples derived from the casting shed. Due to its location and substantial size, it was assumed to be part of the massive steam-boiler chimney structure shown in the 1860 photograph of the site.

Encouraged by this furnace-period discovery, we decided to take a look at any associated subsurface remains, so a 6.6-by-12-foot test pit was excavated directly adjacent to the stone wall. As shown in figs. 4.3 and 4.7, a number of features were exposed. The more substantial ones consisted of a dished brick floor, composed entirely of half-bricks, that formed a sloping basin; the remnant of an abutting brick wall on the south of the basin that had been robbed of most of its bricks; and, to the west, a trench disturbance that also was believed to be a robber's trench. This last feature was found to contain enigmatic *postmolds* (the filled holes left behind after posts has been removed from the ground or rotted in place) and wood remains. The "robbing" attribution given to the features refers to the common practice of retrieving and recycling useful building materials. Most likely this occurred during or just after the demolition of the site.

A 16.5-by-3.3-foot test trench was cut across the northeast end of the basement. The miscellaneous brick features shown in figure 4.3 were exposed at different depths in this unit. Due to their orientation, all are believed to be associated with the furnace period, although their precise function is unknown. Also encountered was the rock ledge—part of the limestone bluff itself—that ultimately supported the archaeological features. Extending the test trench outside the basement foundation to the large stone retaining wall produced only a deep deposit of clay fill and large stone debris.

During 1983 the second field-school reexcavated the original 6.6-by-12-foot test pit in the basement area and extended it over to the test trench. Found in the extension was a large pile of brick debris and rubble overlying a second brickbat-lined basin. Although the second basin was in poor condition, its similar depth and size clearly indicated that it was a companion to the first. A square-shaped brick superstructure of some kind, represented by foundation elements under the pile of debris, was present between the two symmetrical basins. Both basins were associated with narrow, brick-lined "slots" situated at right angles to each other.

We can only speculate about the particular functions of these features, but their placement strongly suggests that they were associated with the steam boiler (refer to fig. 4.3). The basins and slots may have been used as ash or cinder pits under a steam-boiler firebox grate, providing access during fueling or cleaning of the engine boiler. Unfortunately, further excavation to completely expose the intriguing boiler-area features was not possible. It would have required cutting into the slope to the south, both a dangerous and an expensive proposition. Having to content ourselves with only a partial glimpse of this section of the site, our attention turned next to the casting shed.

The Casting Shed

Of all the activity areas present at the site, the casting shed one was the most obvious and well known prior to excavation. The wall foundations to the shed were substantial and relatively easy to locate. They were found to be in almost perfect alignment with the cardinal directions, providing a convenient and useful spatial clue for interpreting other less well-known features. In addition to tracing out the dimensions of the north and west walls, however, we also excavated several pits within the casting shed proper in order to discover what, if anything, remained of casting-related features.

As mentioned earlier, the casting shed had first been exposed by run-off from a storm sewer, which had cut an erosional gully through the approximate center of the structure. This gully exited the shed through the north doorway shown in the 1858 *Harper's* woodcut. Besides exposing the foundations to the shed's doorway and north wall, the gully also revealed a layer of what appeared to be casting sand that occurred between other strata. Thus, the first archaeological task for this area was to vertically trim the sides of the gully to better identify and record stratigraphic information. Inside the casting shed, inspection of the cleaned east profile showed successive heavy layering of tan sand and black sand with charcoal, coal, and cinder inclusions. The tan layer was easily identifiable as casting sand, and the darker lenses in it, complete with debris from both wood and coal/coke combustion (charcoal and cinder, respectively), were thought to represent specific casting events. The casting layer appeared over several strata of construction fill composed of limestone, dolomite, and orange clay. Prominent at the top of the tan-and-

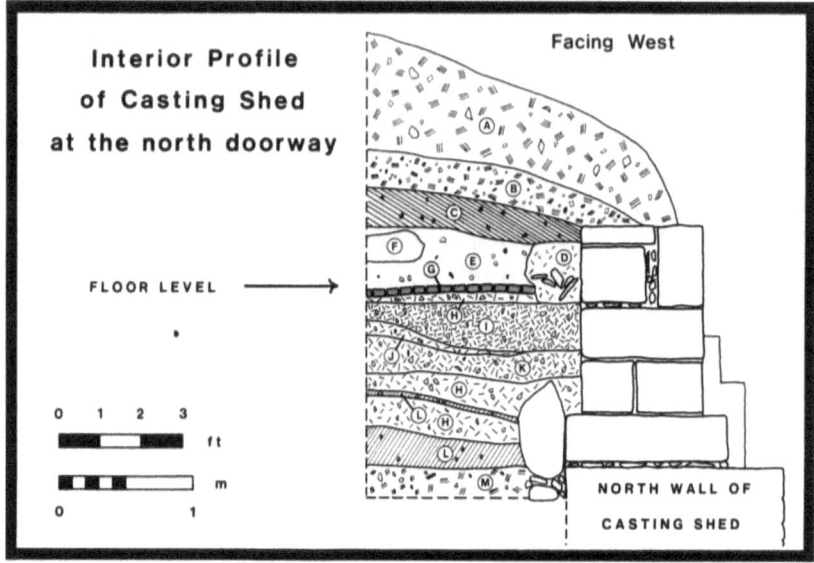

Fig. 4.8. Stratigraphy inside the casting shed. This sectional view of the soil layers inside the north doorway of the casting shed reveals construction, use, and demolition events associated with the furnace. Several layers of fill (labeled H through M) formed the floor of the casting shed. Layers deposited during the furnace operation are labeled D through G, and include a brick floor (G). Deposited after the abandonment of the site are modern overburden layers, A, B, and C.

black lensing was a large deposit of a white chalky substance, later identified as lime. When recalling Swank's assertion that the kiln had been reused by Union troops to produce lime, this deposit takes on considerable historical and interpretive significance.

The interior western profile from the erosional gully provided additional information. Besides the numerous horizontal construction fill layers (labeled H through M) shown in figure 4.8, a brick pad appears underneath the casting stratum. Three distinct layers of redeposited fill appeared above the casting level, and these represent the more recent, postfurnace period of the site's history. The brick pad or strip, which measured a brick and one-half wide (one foot) by five feet long, was perpendicular to the north wall of the casting shed. Although its full extent was not determined, it may have functioned as a walkway for workers or a wheelbarrow runway adjacent to the sow or pig molds. Either later additions of casting sand covered it up, or it was purposely buried beneath the casting sand for subsurface support.

A second, more enigmatic brick feature was discovered in the southwest quadrant of the shed. Figure 4.3 shows the location of the 8.25-by-10.75-foot pit excavated in this area during the 1983 field season. After encountering the obligatory layers of modern fill, the distinctive brown sand that marks the presence of the casting layer was revealed. Just underneath the casting stratum was a remnant of a rectangular brick foundation. Three vertical courses of brick were noted in the best-preserved section of the feature, with the bottom course resting on construction fill at about a foot below the casting sand. The portion of the rectangle that was uncovered had interior measurements of 2.8 feet east-west by 1.8 feet north-south; it was on the same alignment as the casting shed. Two suggestions that would explain the function of this feature can be offered, but neither has documentary correlates. The brick rectangle may have simply formed the foundation for machinery used in the shed during the early furnace period. Later, the machine was removed, the upper courses of the foundation robbed, and a layer of casting sand was distributed over the foundation base, leaving a curious feature for archaeologists to discover and ponder 125 years later.

Instead of serving as a foundation for machinery, the brick rectangle may have supported some kind of large casting mold. Of course, this interpretation assumes that direct casting was carried out at Bluff Furnace, and this is something for which documentary evidence is lacking. Additional supporting archaeological data is also scant: as seen in the next chapter, artifacts that would support a direct casting function (exclusive of pig-iron production) are conspicuous by their absence. At any rate, if this direct-casting scenario did occur, it too must have been early in the life of the furnace, as indicated by the stratigraphic relationship of the foundation below the undisturbed casting stratum.

Two discrete test pits were also excavated in the "back" or southern portion of the casting shed. One, a narrow trench dug adjacent to the large modern retaining wall, was in the extreme southeast corner, and functionally it may be more correct to label this location as occurring within the furnace area. The trench first exposed three modern, knee-high cement-and-stone retaining walls, demonstrating that the terracing of this slope was an ongoing process. At a deeper level, two furnace-period features were discovered. One was the edge of a stone wall of dressed limestone that ran roughly east-west, although it was not exactly in alignment with the casting shed orientation. The other feature appeared to be a brick footing that ran roughly north-south. Very little of either feature could be safely exposed in the deeply dug trench.

Using the trackhoe, a second, irregularly shaped pit was dug about 16 feet to the southwest of the trench. This excavation was contiguous with Brown's "Test Pit A" dug in 1977. The outline of his earlier, backfilled pit was clearly identified. At the bottom of the unit, about 4.5 feet below the surface, another limestone wall foundation was revealed; it may have been part of a doorway to the back of the shed. This wall probably connected to the foundation to the northeast, although it was not perpendicular to it. Other than noting the presence of the limestone-and-brick foundations in the back of the casting shed, there is little that we can say about them. The 1864 photograph (fig. 3.9) shows wall sections in the same general vicinity as these foundations, but they apparently are not on the same orientations as the archaeological examples. Only further excavation and fuller exposure of these deeply buried remains can clarify their structural functions and associations. Both excavation units had to be backfilled in order to bring heavy machinery into the furnace area to the northeast.

Determining the size of the casting shed was the focus of considerable interest and effort. Once the north doorway area had been cleaned and mapped, it was a simple matter to follow out the casting shed wall by cutting along the top of the foundation, using the backhoe and hand labor. The eastern portion of the dressed-stone wall extended a little less than twenty-five feet from the doorway to the furnace foundation. On the west, the wall extended a similar distance to the northwest corner of the casting shed. At the approximate center of this western section the wall had been breached to allow placement of a modern cast-iron pipe. This pipe was installed by the Tennessee Department of Transportation to route runoff water from Riverfront Parkway.

The exterior northwest corner of the casting shed was broken away, as shown in figure 4.3. On the interior we discovered a square brick pad.[7] An educated guess as to the function of this feature is that it served as a footing for columns supporting the roof of the casting shed. In viewing the 1860 photograph (fig. 3.7), note that no timber projects over the top of the walls. This means the roof surface was fairly flat and certainly recessed. Interior piers, resting on pads like the one in the northwest corner, may have been employed to support the flat, recessed roof.

Again using the 1860 photograph, which clearly shows the west walls, two windows, and doorway of the shed, an attempt was made to archaeologically locate the latter feature. In the interest of saving time and labor, a backhoe- and shovel-dug test pit was placed at the expected location of the

Fig. 4.9. West doorway of the casting shed. At the center is the north threshold of the west doorway, flanked on the right by a modern concrete block wall associated with twentieth-century activity at the site.

doorway threshold. Although somewhat disturbed (a modern concrete-block wall is adjacent to it), the north edge of the doorway was easily identified, right where it was supposed to be (fig. 4.9).

No attempt was made to expose the south edge of the doorway, nor to uncover the southeast corner of the shed. The presence of a considerable amount of unstable fill associated with the relocated storm sewer, as well as the sewer elements themselves, prevented safe excavation of these features. Instead, using archaeologically known distances, we were able to scale off the length of the west wall of the casting shed from the 1860 photograph. The extrapolated length of the west wall, and the location of

Fig. 4.10. Two furnace bricks. These specially-formed refractory bricks lined the interior of Bluff Furnace. Despite their large size, the bricks were surprisingly fragile. The composition of the clay made them stable at high temperatures and chemically nonreactive with the melting burden of the furnace.

the doorway in it, are presented in figure 4.3; they measure about 50 feet and 9.5 feet, respectively. The corresponding dimensions for the north wall and doorway, which were directly measured, are 60 feet and 10 feet.

The Furnace Base

Although it is correct to think of Bluff Furnace as a complex whole composed of several functionally integrated parts, the critical part was the furnace. Much of the human activity that occurred at the site revolved around charging the furnace, putting it into blast, maintaining the blast, and tapping the molten iron. Charging and blast maintenance were continuous, round-the-clock procedures. Thus, the furnace was the primary focus of activity at the site, both for the furnacemen who worked there in the 1850s and the archaeologists who would come later.

While it was sufficient to sample the casting-shed and engine areas, the excavation approach for the furnace was one of total coverage, exposing as much of the area as possible. Keep in mind that the preexcavation

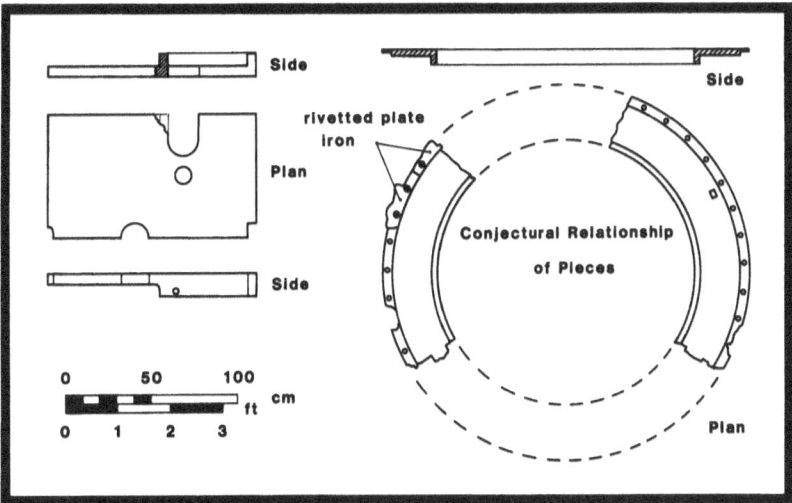

Fig. 4.11. Heavy structural cast-iron pieces from the furnace area. These extremely heavy iron fragments escaped the scrap salvors at the site, being buried deep in the debris near the furnace base. The circular pieces at right were probably part of the tunnel head or throat of the cupola furnace.

model of the site predicted that foundations and other features of two distinct types of furnaces would be present and possibly superimposed on each other: the traditional charcoal-fueled limestone stack, built in 1854, and the short-lived coke-fired cupola furnace that was raised in 1859. If any remnants of the earlier furnace were present, they would probably occur beneath those of the cupola, and both furnaces would be covered by demolition fills and other, later deposits. Clearly, the complex depositional history of the site required both intensive and extensive excavation methods.

An elaborate procedure was employed to remove the three to ten feet of spoil over the furnace base. Initially a backhoe and a front-end loader were used to loosen the fill and move it to the southwest corner of the casting shed. There it was scooped up with the trackhoe that was sitting at the engine-area level. Finally, the fill was transferred to a dumptruck and hauled away. Once the furnace-period depth was reached, the heavy machinery was replaced with shovel and trowel excavation techniques.

During removal of the spoil in and around the ravine, and in clearing the furnace area of demolition fill, numerous whole and partial furnace bricks were encountered. The whole bricks were recovered and stored for

Fig. 4.12. Initial exposure of the cupola base. Our first view of the base, abandoned after the Civil War, was surprising. We encountered the salamander fused into the hearth of the furnace. Beneath the scale is a cavity left in the salamander by the tuyere next to it. When the mass of coke, iron, and slag solidified in November 1860, only one tuyere was still in blast. Displaced furnace bricks ringed the salamander.

possible use in the reconstruction and interpretation of the furnace. Although there was a good deal of variation in firebrick dimensions and form, a "typical" example of a furnace firebrick measured a foot long, four inches thick, and tapered from about seven inches at one end to five and one-half inches at the other. As with old-fashioned well bricks, this tapering ensured that the bricks were self-locking once they were laid in the furnace stack (fig. 4.10). "Nonstandard" firebricks probably reflect variation in the form of the furnace lining as well as trimming of bricks to conform to furnace fittings. Many examples bore direct evidence of

Fig. 4.13. The cupola base after excavation. This view of the exposed base looks straight into the throat of one of the tuyeres left fused into the salamander. Sandstone blocks formed the structure of the hearth and were held together by the wrought-iron restraining band at the bottom.

their function: slag and other furnace materials were found adhering to their tapered (that is, interior) faces.

In addition to furnace bricks, several large pieces of iron were recovered from the demolition fill over the furnace base. Three are illustrated in figure 4.11. The two large fragments of a concentric ring plate or some other kind of structural element may be part of the tunnel head of the stack. The inside diameter of these cast-iron pieces is 4.86 feet, which probably represents the stack diameter at the tunnel head. It does *not* represent the diameter at the base of the furnace, which was determined to be in excess of eight feet, as discussed below. The third fragment is a heavy structural cast-iron piece of unknown function. Other miscellaneous artifacts from various proveniences are discussed in chapter 5.

The principal and most exciting discovery in the furnace area was the sandstone hearth at the base of the cupola. When considering how little of the site's features remained in the 1864 photograph (fig. 3.9), and the potential sources of destruction threatening these scant remains over the intervening century, it is only a fortunate accident of preservation that

Fig. 4.14. Overview of the cupola base. The two foundations of Bluff Furnace are seen in this overhead photograph taken from the charging deck. The hexagonal footing surrounds the base of the cupola furnace and sits atop the angling walls of the charcoal furnace.

any substantial features were still extant at Bluff Furnace. As seen in the photograph, the lower portion of the remaining cupola stack—the area below the supporting arches—consists of a series of cylindrical sections arranged in the form of an inverted cone. The sandstone ring that was found was the lowest of these cylindrical sections shown in the photograph.

When first encountered—that is, when the highest surviving portion of the furnace base was reached—we were pleasantly surprised to find a salamander (a conglomerate of fused iron, coke, and flux) surrounded by furnace brick and sandstone fragments (fig. 4.12). Apparently this exceedingly cumbersome and heavy feature had been ignored when the bottom of the furnace shown in the 1864 photograph was finally demolished. We removed an upper portion of the salamander down to solid iron, and when the soil matrix around the hearth was cleared, the circular form of the cupola base was revealed. The lining of the hearth was formed from large sandstone blocks bound at the base with an iron restraining band

Fig. 4.15. The forehearth area. The large stone at left may represent an attempt to repair a rupture in the lining. Covering the upper surface of the stone is a tongue of the salamander. The dark soil leading to the front of the furnace is a casting channel.

about ten inches high (fig. 4.13). The outside diameter of the hearth was 8.6 feet. Immediately above the sandstone blocks were the remains of two metallic tuyeres, one of which is clearly visible in figs. 4.12 and 4.13.

The spacing of the two tuyeres on the perimeter of the hearth indicates that the cupola had originally been furnished with five tuyeres on sixty-degree radials. The front of the hearth, where the molten iron was tapped, probably was not furnished with a tuyere, as was standard practice at the time. The best-preserved and most complete of the two tuyeres was conical in shape and had its nose embedded in the salamander. Unlike the water-cooled tuyere in figure 2.12, in this example the pipe delivering water to the tuyere was carried to the nose through a separate conduit molded into the side of the main section; the egress line spiraled out through the shell of the tuyere. The mouth of the water jacket was one-half foot in diameter. The tuyere's full length is unknown due to the position of the nose within the salamander, but at least a foot and a half was measured.

As can be seen in figure 4.12, the force of the blast through the tuyere had formed a large bubble in the mass of partially and completely reduced iron, graphite, coke, and slag making up the salamander. When the contents of the furnace slowly cooled, the cavity became frozen in time and space. Swank's commentary was apparently both literal and figurative when he noted that in 1860 "the furnace soon chilled."

Once the matrix around the hearth was completely removed, the underlying foundation shown in figure 4.14 was exposed. This consisted of a three-foot-wide stone footing in the form of a hexagon, with the circular hearth in the center. The western side of the hexagon, that is, the one closest to the casting shed, was breached, and instead of heavy ashlar limestone blocks, it consisted of small sandstone slabs set below the level of the other foundation courses (refer also to fig. 4.3). This area was the "front" of the hearth where the molten iron was tapped, and it was over these sandstone slabs that casting sand had been drawn up to the furnace base during its operation. At each corner of the hexagonal footing were remnants of metal plates that had originally anchored the feet of cast-iron pillars supporting the cupola. This is exactly what we would expect from seeing the 1864 photograph of Bluff Furnace, which shows four of the six iron pillars (the other two are obscured by the furnace base itself).

A small brick pad, one and one-half bricks wide, appeared behind one of the corner plates. This pad apparently had butted against the inside angle of the corner post, perhaps bracing a broken pillar. As is clearly seen in figure 4.14, the space between the circular hearth and the surrounding hexagonal footing had been nogged with whole and half-bricks. At three locations within this brickwork, flat metal plates were noted, perhaps having served as pads for vertical supports of the overhead blast pipes.

Much intriguing information was uncovered at the front of the hearth, called the *forehearth*, where in traditional furnaces the damstone and timpstone were located. To the north of the forehearth was a large limestone block, situated outside but abutting the edge of the circular hearth, while to the south remnants of an irregular brick feature appeared; both features are clearly visible in figure 4.15. A half-brick that was part of this construction was marked with impressed lettering that read "SOUT . . ." and "MAN . . . ." [8] Between the brickwork and the limestone block was a narrow channel filled with congealed slag.

One of the most curious attributes of the salamander was its form: while the hearth that held it was circular, the salamander itself was not. Its oblong shape, together with the irregular stone- and brickwork present at the forehearth, led to an important conclusion about how it came to be there in the first place. When the front half of the brittle upper portion of the salamander was removed, a more dense metallic layer was discovered beneath it. This would have consisted primarily of molten or pasty iron when the furnace was in blast. The salamander was confined more or less within the limits of the circular inner hearth walls—except for the front, where the metallic portion projected outward to the west. The projection extended beyond the limit of the hearth lining to the top of the large limestone block at the hearth front.

What are we to make of this irregular arrangement? The most feasible scenario that can be offered is that the lining of the hearth failed, releasing the then-molten salamander through the hearth wall until it cooled and solidified. The large limestone block at the front of the hearth represents a desperate attempt to revet the hearth during the displacement of the lining. The repair attempt failed, as indicated by the tip of the salamander projecting over the limestone block.

In other words, during Bluff Furnace's final blast, the front lining of the hearth at the tuyere level was displaced outward. An attempt was made to halt the displacement of the lining by revetting the front of the hearth with stonework. The emergency repairs failed, and the salamander pushed its way past the hearth lining onto the limestone block, where it cooled upon contact with the air. Only one tuyere bubble was present in the salamander, indicating that the other four were not operable during the emergency. If the lining failure had damaged the water pipes servicing the tuyeres, the results would have been much more dramatic than the formation of an iron, coke, cinder, and slag salamander. Water for the tuyeres would have been violently released in the form of scalding steam.

The archaeological data documented that the inwall of the furnace at the level of the tuyeres had failed near the front of the hearth. An examination of the surviving sandstone blocks reveals a possible design flaw: the blocks coursed at the same level, meaning there was a continuous

Fig. 4.16. Brick wall and tuyere pipes. The brick foundation south of the furnace base probably served to support the blast machine and shows many modifications indicating repairs or remodeling. The two curved pipes in front of the foundation were exhaust-water lines from the water-cooled tuyeres of the coke furnace and fed into a wooden trough, the remnants of which were exposed along the face of the brick wall.

seam between one course of hearth blocks and the one above. The tuyeres entered through notches cut in the courses above the salamander. These courses were absent at the time of our excavation and were probably removed when the furnace was dismantled. Each course was revetted with metal restraining bands such as the one exposed near the furnace base. The underside of each outward step was probably supported by heavy cast-iron bed plates, called *mantles*.

Once the hearth inwall had failed, displacing a sandstone block outward, the mantle plates supporting the overhanging courses of blocks may have also ruptured, ripping open tuyere feeder pipes and water-cooling lines. When the salamander congealed, at least one tuyere was left in full blast, with the result that a bubble was cast into the salamander. Bluff Furnace had not been "blown out" in the normal manner. There had been a serious accident in the hearth. The failure may have been the result of a structural weakness in the masonry or ironwork or some human failure in the management of the blast. The accident was costly in any case.

Directly in front of the hearth was the tan sand fill that made up the floor of the casting shed. A dark linear soil stain, running east-west, is what remains of the main casting channel leading from the damstone to the mold beds in the shed. A decayed vertical wooden post was present on each side of the channel, in addition to a vertically set iron bolt on the north side. Linear east-west soil stains were present beneath the upper surface of the casting sand and were associated with the posts and bolt. These elements are believed to be remnants of a wooden stake and plank form that contained and supported sand around the primary casting channel connecting the hearth with the casting beds; similar arrangements are illustrated in at least one documentary source.[9] Using the transit and stadia rod, the vertical elevation of the channel at the forehearth was measured and compared to the elevation of the casting sand layer recorded in the profile of the erosional ravine described earlier. A difference of just over five feet was calculated between them. Apparently this drop in elevation between the forehearth and pig mold beds was necessary to help propel the molten iron down to the most distant beds in the casting shed.

When the coke furnace was in operation, a working floor of dense, packed coal cinder and clinker surrounded the hearth base and covered the hexagonal footing. The presence of this floor was a fortuitous discovery because it marked the boundary between the earlier charcoal- and the later coke-period remains. Anything sealed beneath it would have been associated with the construction, operation, or demolition of the charcoal furnace.

Another important discovery was made just south of the cupola base. Oriented to the cardinal directions were the north face and part of the west face of a substantial brick wall. The wall's north face bore evidence of extensive modification to the brickwork (fig. 4.16). This structure may have been a platform upon which the furnace's blast machine rested (the heating oven appears to have been south and west of the furnace base). No diagnostic artifacts or construction characteristics were noted that would confirm this, however. The brick platform does not appear in the 1864 photograph, having been covered by demolition or erosional fill.

We are on somewhat firmer ground in our functional interpretation of the decayed remains of a wooden box or trough uncovered next to the brick wall. It measured 9.2 by 2.4 feet and was constructed of planks nearly two inches thick. A single iron tie rod was found *in situ* on the east end of the box; it apparently served to bind the sides of the box together. The bottom boards rested on three eight-inch-wide joists set into the surface of the cinder floor, which establishes that the box was contemporaneous with the cupola furnace.

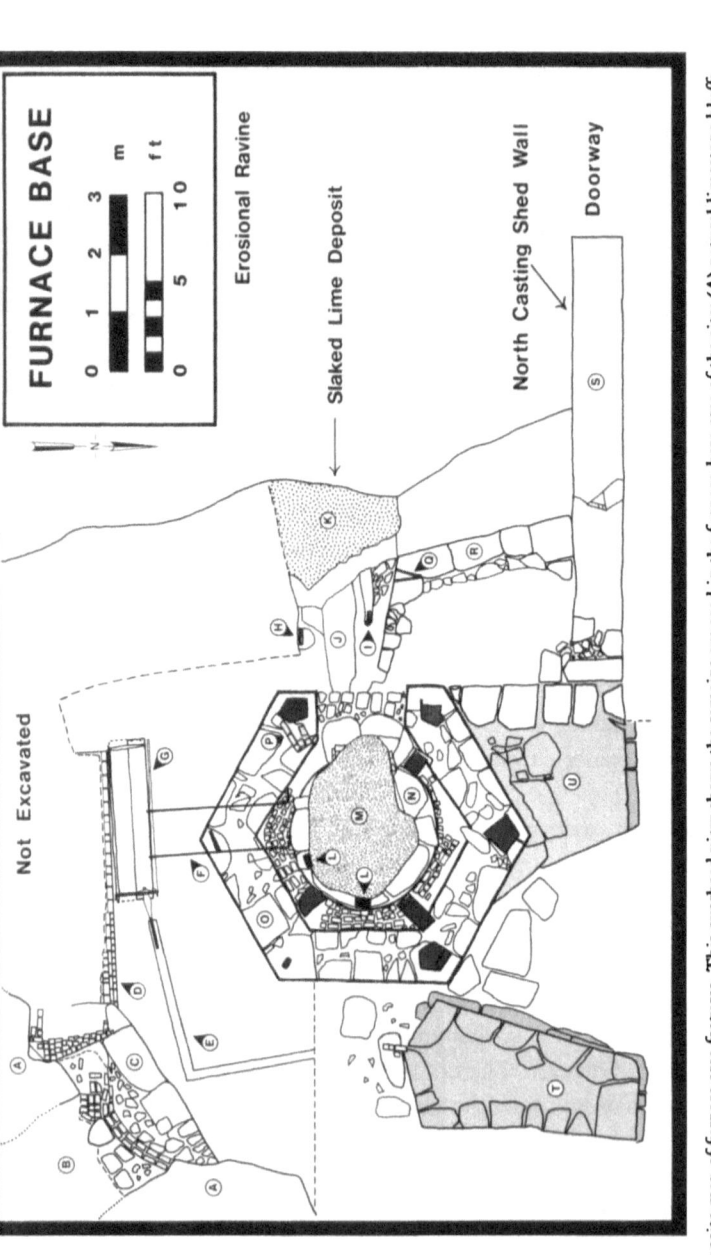

Fig. 4.17. Composite map of furnace area features. This overhead view shows the remains exposed in the furnace-base area of the site: (A) natural limestone bluff; (B) crevice in the cliff face; (C) mortared stone pad supporting curved brickwork sealing the crevice; (D) brick foundation supporting the blast machine; (E) egress trough for tuyere wastewater; (F) iron pipes venting wastewater from tuyeres on south side of furnace; (G) wooden trough into which hot wastewater was vented from tuyeres; (H) and (I) iron bolts associated with liner of casting channel from furnace hearth to casting shed; (J) casting channel; (K) slaked lime deposit from Civil War period; (L) water-cooled tuyeres fused into salamander; (M) sandstone blocks forming outer shell of hearth; (O) limestone foundation of cupola furnace; (P) mortared brick pad associated with cupola furnace; (Q) masonry anchor mortared into limestone pad (R) which served as working surface at the forehearth; (S) north casting-shed wall; (T) and (U) foundations of the charcoal furnace, framing the north tuyere arch.

Three iron pipes associated with the box provided important clues about the box's function. A large pipe egressed from the east end of the box, emptying into a smaller wood-lined trench set into the working floor. As shown in figure 4.17, the trench first ran east and then angled north around the furnace foundation. Two smaller wrought-iron pipes, partially buried beneath the cinder floor, connected the box and the furnace. They both were bent in a vertical S-curve, entering the box fourteen to fifteen inches above its base (see fig. 4.16). At the furnace end they turned toward the west, with the easternmost one angling up toward the vertical. In all likelihood these pipes, which had remnants of some kind of insulating materials adhering to them, collected hot wastewater (probably in the form of condensed steam) from the two closest water-cooled tuyeres and vented it into the wooden box. Why not simply empty the water directly into the drain? A possible answer is indicated by the contents of the box. At the bottom, overlying the decayed planking, we found remnants of congealed slag, with many examples bearing smooth impressions that might have resulted from tools. We believe that the tuyere wastewater was collected in a shallow quenching box provided for tools used to work the furnace.

No other buried pipes were found around the furnace base, indicating that both blast pipes and tuyere feeder pipes were suspended around the furnace base above the working floor. A stone apron or pad was exposed to the northwest of the hearth. Connected to the casting shed's north wall, this feature possessed a carefully laid border of limestone from which we recovered a wrought-iron masonry anchor (refer to fig. 5.3 in the next chapter). The top of the apron was either composed entirely of cinder, or its brick or stone surface had been robbed away. The apron evidently provided a working surface from which furnacemen attended the hearth, removing waste slag as well as tapping the molten iron.

Numerous samples of coke, slag, and other waste by-products of the furnace operation were recovered from above the cinder floor, so all are attributable to the coke-fired period. After mapping and photographing the above-floor features, this cinder surface was removed with shovel and trowel, exposing the underlying foundations of the earlier charcoal-fired furnace. Amidst the debris we encountered a large pig-iron bar. Due to its presence beneath the cupola furnace working floor, it must be associated with the charcoal furnace production (this specimen is discussed further in chapter 5).

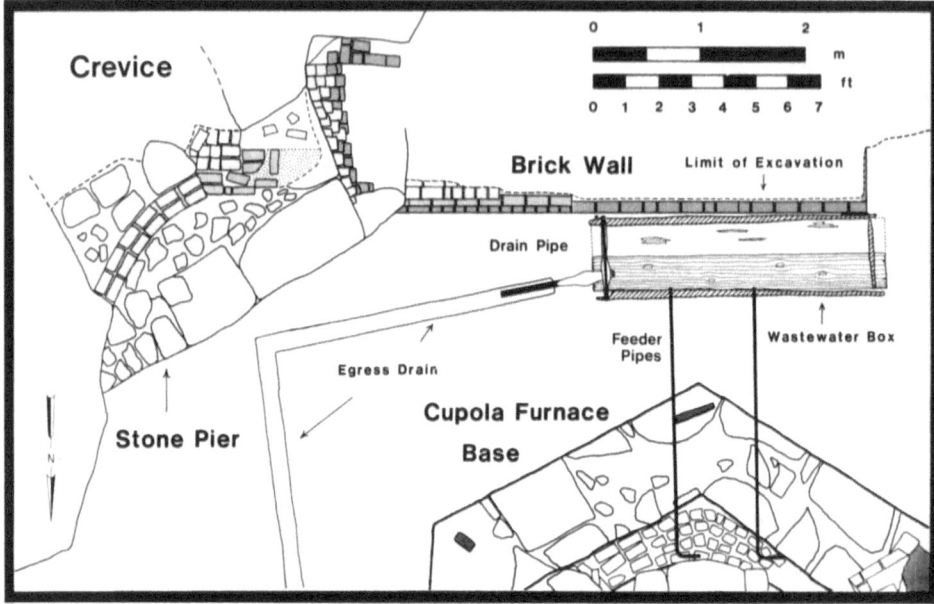

Fig. 4.18. Details of features south of the furnace base.

Figure 4.17 shows the foundation elements from the first furnace that were revealed after the cupola working floor was removed. When it was dismantled in 1859, only the outline of the north tuyere arch was left intact, and this surviving footprint is what is illustrated in the site map. The coke furnace was centered about six feet west of the center of the original furnace. The charcoal furnace had at least one tuyere and very likely two, judging from period practice. The north wall footing of the earlier stack spanned at least twenty-six feet at its base, with a tuyere arch half as wide. South of the cupola hearth, little evidence of the charcoal furnace survived. Both foundations rested on horizontally bedded ledges of limestone geologically connected to the bluff formation itself. Drill holes on the west face of the bluff attest to the amount of effort expended on shaping the bluff to fit the furnaces. A substantial amount of stone had to be sheared away to create a shelf for the furnace base and a vertical face up to the charging deck of the plant.

To the southeast of the furnace base a massive stone pier was obliquely set so as to connect the west face of the rock bluff with the east end of the brick wall described earlier. Extending a little over eight feet above the furnace base, the top of the pier was situated at the bottom of a vertical

The Archaeology of Bluff Furnace / 125

Fig. 4.19. Counterweight shaft at the base of the bluff. At least during the charcoal-fired period at Bluff Furnace, raw materials arriving at the plant by barge were hauled to the furnace top on an inclined ramp from the river. This shallow groove in the bluff face was probably a shaft for a counterweight compensating for the incline carts. Square notches bracket either side of the pipe chase.

crevice in the bluff face (this crevice is visible in the 1864 photograph). Resting on the top of the platform were two brick constructions. The one on the west (refer to figs. 4.5, 4.17, and 4.18) was probably a higher remnant of the brick wall or foundation that begins on the furnace base. The other eastern feature was a short, curved retaining wall, probably built to retain soil in the vertical crevice. Also found, adjacent to the stone pier, was a mass of brick filler. Along with the upper remnant brick construction, the discovery of this feature gives credence to the suggestion made earlier that a substantial brick wall, of considerable height, served to support the blast machine or hot-blast stove.

Not investigated by us was the lower section of the bluff adjacent to the river. There Brown had identified a "pipe chase" in the face of the bluff, formed by drilling and blasting. As seen in figure 4.19, drill scars and iron anchor bolts embedded in the limestone conform to those appearing in the upper section of the site. Presumably this man-made alteration to the bluff face served as a counterweight shaft for hauling up materials to the charging deck from the river. The prominent ramp shown in the 1858 *Harper's* illustration (fig. 3.5) was most likely used, along with a steam-driven winch attached to cables and a counterweight, to move iron ore, flux, and charcoal taken from river barges to the charging deck.

The Slag Pile

Establishing the presence of the slag deposit outside the casting shed was a fairly straightforward proposition. Brown had noted a slag deposit in his preliminary excavations, and while removing modern fill with the backhoe we also encountered a concentration of slag. More definitive information on its presence was derived from excavation and stratigraphic analysis.

Outside the north casting-shed door the erosional ravine had made a deep south-to-north cut. This was enlarged upon and deepened with a combination of backhoe and extensive hand-labor. When the east profile of this cut was troweled down, a highly lensed soil stratigraphy was revealed. Clearly discernable in this sequence was a thick layer of slag that was apparently deliberately deposited in a shallow trench (layer H in fig. 4.20). The sharp vertical cut on the southern edge of this trench indicates that it was filled relatively quickly after it was dug: such an acute angle would have eroded away into a more gentle slope had it been exposed to the elements for even a short period of time. Limitations on time and resources prevented us from learning the full spatial extent of the slag deposit. We do know that it did not extend as far as the west profile on the other side of the ravine. Samples of the slag were collected from this distinct archaeological context for laboratory analysis.

Fill layers deposited during the construction of the casting shed are prominent at the base of the profile, and they demonstrate the sharply sloping surface of the site during the furnace period. Piquing our interest were what appeared to be two broad depositional events, represented by layers of tan sand and black sand with cinder lenses and separated by a

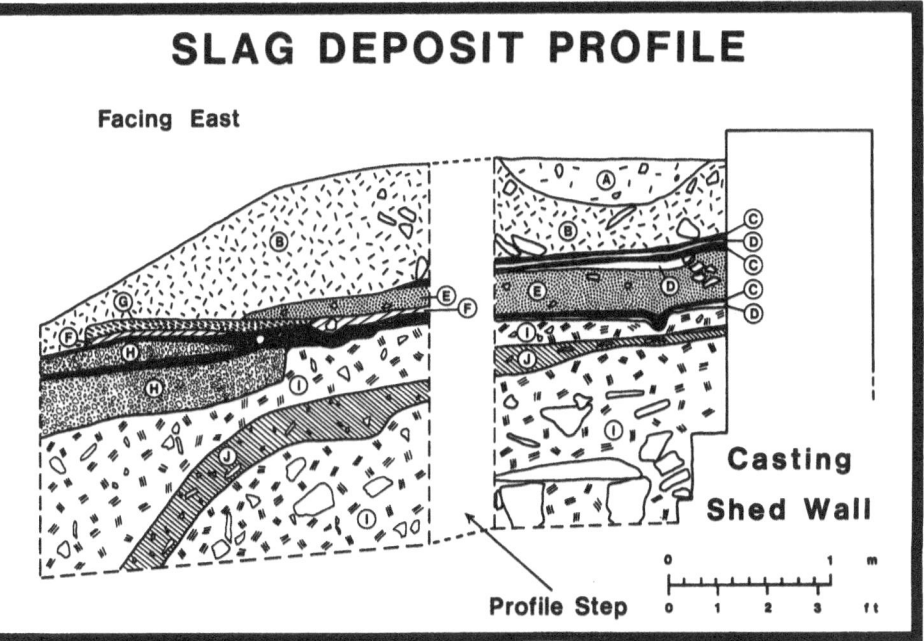

Fig. 4.20. Soil profile north of the casting shed. Like the interior stratigraphy of the casting shed, the exterior of the structure was built up with layers associated with the construction, use, and abandonment of the furnace: (A) backhoe disturbance; (B) fill deposited after the furnace was closed; (C) black soil with charcoal inclusions, spoil dirt from the casting operation; (D) discarded casting sand; (E) brown silty soil with charcoal and small brick fragments; (G) red silty soil; (E), (F), and (G) possibly associated with furnace remodeling; (H) dark green to black slag deposit; (I) red and tan clay with limestone debris, primary construction fill; (J) reddish-brown clay and limestone gravel, primary construction fill.

thick zone of brown silty soil and brick debris. It is tempting to associate this 1-2-3 sequence with the documented initial charcoal-fired period, the demolition and conversion effort of 1859, and the final coke-fired period of operation, respectively. Without additional supporting data, however, this temptation must be resisted.

What the profile adjacent to the casting shed does provide is a sense of the impressive scale of effort needed to construct the furnace. The casting-shed foundations were excavated to a depth of ten feet below the level of the sill of the north door without ever reaching the basal course. Together with the acute slope of the construction fills shown in the profile, this indicates the tremendous investment of materials and labor that were required to transform the bluff into a viable industrial site.

The fragile furnace base could not be left exposed for an indefinite period; the soft sandstone hearth lining would deteriorate rapidly, and vandals or souvenir collectors would surely attempt to extricate the tuyere fused into the salamander. Exposure of the cupola base, mapping, photography, and artifact retrieval were all accomplished in one long day. To ensure that the most important remains at the site remained protected from the elements and vandalism, we used brick and stone debris to construct a cairn around the hearth. After lining the salamander with heavy plastic and covering the surface of the stack foundation, we raised low brick and stone walls from the hexagonal cupola foundation. The interior of the cairn was filled with casting sand brought in, appropriately enough, from a local foundry. The earthen mound covered the base of the cupola stack and sealed the archaeological remains from the weather.

The cupola mound and other areas of the site were seeded with grass in order to stabilize the ground surfaces for the winter ahead. The field was abandoned for the lab, where the research started at the site could be finished.

Archaeologists are accustomed to viewing fragmentary remains and attempting to reconstruct what the original feature or object must have looked like. Some of the furnace remains were not very intelligible, in part because of the dismantlement of the furnace machinery, the subsequent deliberate demolition of the site, and the robbing of construction materials and scrap iron. Robbing probably applied to the removal and salvage of any and all usable materials such as bricks, cut-stone slabs, timber, nuts and bolts, etc. Even broken metal items could be sold for scrap to foundries. It is axiomatic that archaeologists often find only what was considered in the past to be good for absolutely nothing.

It was apparent that the construction of the furnace plant had involved a great deal of labor and engineering. The overall geology of the bluff consisted of bedded limestone layers dipping to the south and west. In a horizontal plan, two terraces were created, one above the other. The furnace base rested on a rock shelf trailing off in a line to the southwest. To create the lowest platform, the north casting-shed wall had been anchored on bedrock paralleling the river and carried to a height sufficient to create a kind of dam. The structure was enclosed on the west and at least partially on the south by walls. Rubble stone and clay subsoil from clearing upslope were filled down behind the north casting-shed wall, creating a level surface for the casting-shed floor. Above this first terrace, the west face of the bluff was trimmed vertically to create a wall against which the furnace stack was situated.

Twenty feet above the furnace base–casting shed terrace, a second terrace was created using the base of a natural limestone shelf. On this terrace, the lower supports of the charging deck were erected, in the process leveling out the uneven contours of the cliff face terrace. From the west bluff face in a line running southwest, a retaining wall had been built to link the upper and lower terraces. The remnants of this buttressed wall are prominent in figure 3.9. This was the basic structure of the site, then: a north-south terrace anchoring the charging-deck and stock-house level of the furnace, a diagonal retaining wall, and an east-west terrace for the furnace base and casting shed.

From these terraces the furnace-plant superstructure was raised. At the highest point in the charging-deck terrace, vertical anchor bolts were placed in the surface of the limestone shelf, anchoring the river side of the charging deck. Along the length of the charging-deck shelf, brick piers were raised to support the timberwork of the deck and its roof. The stock house and rooms for winching equipment were added. The retaining wall buttressed these constructions and formed a vertical wall at the southeast corner of the furnace–casting-shed area. The casting-shed walls were raised above floor level in native stone, perhaps cut from the bluff itself or nearby. The stone stack of the charcoal furnace rose against the west face of the bluff, perhaps even mortared into it for extra strength.

The charcoal furnace had two tuyeres, on the north and south side of the stack, and the forehearth was on the west. The hot-blast stove was on the top of the stack to the west of the larger draft stack. A cast-iron downcomer pipe descended along the north furnace face. Whether the high smokestack shown in the 1860 photograph (fig. 3.7) was constructed in 1854 or in 1859–60 is a moot point.

When converted to burn coke, the old charcoal stack was demolished, leaving the outline of the north tuyere arch foundations in the ground. Building above this point, the hexagonal base of the coke cupola stack was erected. The large sandstone blocks forming the floor and lower walls of the hearth were custom fitted together and then bound with a riveted sheet-iron band. Brick nogged the space between the hearth stones and the cupola foundation. Notches for the tuyeres were cut in the hearth inwalls, and the stack rose toward the boshes in a series of stacked cylindrical sections supported by cast-iron bed plates. At the level of the boshes was a structural cast-iron rim plate anchored into the stonework.

The plate was supported by a series of six cast-iron pillars with arches between. This structure, the mantle plate, supported the weight of the remainder of the stack, and the bustle pipe feeding the tuyeres was suspended from this structural plate. Above this base plate rose the cylindrical shell of the stack consisting of riveted iron plates surrounding the fragile firebrick lining of the stack. At the furnace top was another cast-iron rim plate structure, this time surrounding the tunnel head or throat of the stack. It is not clear whether the tunnel head was open and the stack charged from the top, or whether doors opened from the east side of the upper stack and charges were admitted from that point.

The charcoal stack had two chimneys at the furnace top, one being a draft stack and the other being the hot-blast stove into which a portion of the waste top gases were vented. This furnace top stove arrangement was replaced by a separate hot-blast stove situated above the roof level of the casting shed. A brick downcomer brought the top gases down toward the level of the casting shed, and through unknown means the gases were conveyed to the hot-blast stove, probably entering from its base. The stove had two small exhaust stacks through which the waste gases were vented after exchanging heat with the fresh blast air. A cast-iron blast pipe containing the preheated blast exited the stove in a line with the furnace stack.

The type of blast machine employed at the furnace and its location remain unknown. The presence of the high draft stack at the south end of the charging deck and the presence of raw coal at the site indicate that a coal-fired boiler was present on the charging deck. It is probable that steam generated at this boiler was piped to a steam engine situated either on the charging deck or in the southeast corner of the casting shed. Certainly the expensive steam engine would have been placed under shelter. The engine drove a blast machine of unknown design, probably situated somewhere under shelter near the furnace base. The pressurized blast was then piped up to the hot-blast stove.

It is fairly typical that as many questions are raised by field research as are answered. After putting the site to bed for the winter, we returned to the lab to count our artifacts and blisters, study our notes, and try to reconstruct the operations of Bluff Furnace. Now the focus of attention was on the furnace-period artifact assemblage recovered during the excavations and the information that these objects provided about specific behavior that took place at the site.

CHAPTER 5

Bluff Furnace Artifacts

Excavations at Bluff Furnace brought to life the historical accounts of the furnace operation and provided intriguing information on the final blast of the furnace. But archaeological research never stops at the nominal level of discovery, that is, the digging up and recording of "old things" in the ground. Fieldwork, while the most exciting and rewarding aspect of the project experience, is only the beginning of research, not the end.

Most of the artifacts we recovered at Bluff Furnace consisted of fragments of piping and duct work, structural plates, and the like. But we also excavated through enormous quantities of raw materials and smelting by-products. These materials were our key to understanding the operation of the furnace because they provided access to events that occurred over 130 years ago. In the same manner that the Bluff Furnace ironmasters had examined slag to determine how the blast was progressing, we hoped to do the same. There were questions about the nineteenth-century smelting operation that could not be answered in the field, questions concerning the quality of iron that was produced, the origin and qualities of the ores and coke, and the efficiency of the smelting process. Only detailed chemical analyses of the ore, coke, and iron samples could address these specific questions. With this in mind, we systematically took samples of coal, coke, iron ore, and slag deposits exposed at the site.

Unfortunately, there is little contemporary archaeological data on nineteenth-century blast furnaces to which to compare our findings. Archaeologist John R. White is one among a handful of researchers who has subjected nineteenth-century blast-furnace materials and by-products to scientific scrutiny.[1] His examination of the Eaton-Hopewell Furnace in Ohio, which operated in the period 1802–8, yielded a surprising and unexpected result: the earliest documented use of mixed coal-and-charcoal blast fuels. The chemical analysis of blast slags from this site revealed their poor ability to remove sulfur from the iron, a factor that White believes contributed to the demise of the furnace operation. White's work demonstrated that modern analytical techniques could provide significant new dimensions to purely historical accounts of iron smelting, a lesson that we took to heart in our work at Bluff Furnace.

Sampling and Blast-Furnace Archaeology

Sampling blast-furnace materials and by-products in a manner that would adequately reflect the functioning of the smelting operation was a problem. Which pieces of coke, which bars of pig iron, and which chunks of slag accurately characterize the "normal" blast? Statisticians refer to *validity* in sampling schemes. That is, do the samples really tell us what we think they are telling us; is there a real connection between the sample and the phenomenon being sampled? Statistical *reliability* is a term referring to the relative confidence we can place in a sample as indicating a larger whole. Statisticians know that, as sample size increases, reliability increases.[2]

The chemical and physical character of iron ores and coals varies considerably within discrete ore and coal beds; there is simply not a uniform, precise composition throughout these geological deposits. Similarly, slag composition may vary in the same manner that the raw materials vary; it is not difficult to imagine that the slag produced when the furnace is first being brought up to full blast differs from that produced when full blast has been achieved. Further, blasts at a single plant differ, as when a new fuel, ore, and flux mixture is tried.

In archaeology we speak of the *context* of an artifact, its place in the temporal-spatial matrix of a site and the assumed behavior that led to the deposition of that artifact. In other words, what does the artifact mean? At Bluff Furnace, the most important relevant archaeological context question was whether this slag/pig iron was from the charcoal-blast period or the coke-blast period.

Fortunately, at the Bluff Furnace site it was sometimes possible to distinguish between deposits associated with the charcoal period and those of the coke period. Only above and below the working floor surrounding the coke hearth were we able to safely distinguish the charcoal- and coke-period deposits. As the slags had been flung down the riverbank north of the furnace, there was very little contextual control. John White sampled ore and fuels lost from the charging deck of the Eaton-Hopewell furnace and made a point of obtaining multiple samples from the upper and lower levels of the deposits, making what was for his research the critical contextual distinction between the charcoal-only period at the Eaton-Hopewell Furnace and the second blast phase, where charcoal and raw coal were mixed.

Unfortunately, archaeological deposits are sometimes not amenable to the collection of multiple samples which, when analyzed, would yield a more reliable assessment of behavior than small samples would. For example, at the Bluff Furnace, our sample of pig-iron bars was limited to four specimens; the small size of this sample was simply beyond our control.

The Artifacts of Industrial Sites

The kinds of debris recovered from industrial archaeological sites are strikingly different from those recovered at prehistoric or historic period sites of the "garden variety." Industrial sites and artifacts are frequently large in size and require special handling. The cumulative weight of the iron Bluff Furnace artifacts was staggering, and several mashed fingers attested to the difficulty in simply handling some of these items. Not surprisingly, very few intact machine parts were recovered from the site; we anticipated this, knowing that the furnace had been nearly completely dismantled in 1863. Doubtless, sometime in the late 1860s, or 1870s, what was left of the iron cupola stack *cum* lime kiln was probably broken up for scrap and remelted in one of Chattanooga's many foundries. Consequently, we did not expect to find any machinery or tools on the site, except in very fragmentary form.

The laboratory and analysis phases of an archaeological research project typically involve four discrete functions: cleaning, artifact inventory and classification, artifact analysis and data synthesis, and report preparation. Cleaning is one of the more mundane things done before starting the analysis of the artifacts. Before the function and significance of an item can be determined, it must be thoroughly cleaned; even a few years in the ground can obscure the form and function of an artifact, particularly if that item is made of iron.

When left unprotected by special coatings and finishes, and not maintained by constant cleaning, metal begins to decay. This is particularly true of ferrous objects. In a general sense, the metal begins to revert to a ferrous oxide form akin to iron ores; there are as many varieties of rust as there are of ore. Through time, some metal artifacts reach a decay equilibrium in the ground; oxidation slows or stalls at a particular stage. Ferrous artifacts, when subjected to ground moisture and contact with soil, rust quickly on the surface. If unchecked, the decay continues layer by layer into the underlying metal. If the item is thick, the accumulation of corrosion on the outer surface of the object might slow or stop further corrosion underneath by preventing oxygen from contacting the uncorroded metal. In this instance, the continued formation of oxides (rusting) slows.

In removing the metal object from the ground, the archaeologist may actually be renewing the process of decay. By removing the insulating soil in which the object was buried, air may again reach the interior of the object and decay begins anew. One of the most tedious but productive laboratory procedures undertaken in the analysis stage of the research is the cleaning of unstable metallic objects. The main process for cleaning and stabilizing ferrous objects is *electrolytic cleaning* or *electrolysis*.

Electrolysis is a form of reduction not unlike smelting; the process involves breaking down ferrous oxides with energy—in this case, electricity. Artifacts are suspended in an electrolytic solution, usually a dilute lye such as sodium hydroxide, and are attached to the negative terminal of a battery charger. A conducting metal plate situated next to the artifacts is attached to the positive terminal. When the charger is turned on, an electrical circuit is created between the positively charged plate and negatively charged artifacts. This induced flow of electrons breaks down the chemical bonds in the ferrous oxides forming the rust, while the stronger chemical bonds in the pure, uncorroded metal remain unchanged. Eventually the rust begins to spall off the underlying metal. Careful hand cleaning with wire brushes speeds up the process.

Once all oxides have been removed the artifact is desalinated in a bath of pure distilled water. In this stage of processing, salts absorbed from the ground are leached out of the metal. Otherwise, salt molecules in the metal would begin to attract moisture, and rusting would begin anew. After the artifacts are dried they are coated first with stabilizing chemicals and then with sealants to prevent contact with the air. A clear acrylic coating is applied as a last step in the conservation process; with this protective layer covering the metal, the object is chemically stable for decades.

Small objects are easily cleaned in electrolysis tanks unless they are made of alloys resistant to the process. Larger objects can be difficult to treat, and industrial archaeologists have the disturbing habit of collecting some very large artifacts. Professional conservators have developed methods to clean even enormous cannons from shipwreck sites, but the process is slow, labor-intensive, and therefore expensive. If the object is not decorated with fine embossing, or otherwise fragile, and no information will be lost, the item can be cleaned by sand blasting. Blown by air pressure through a control nozzle, fine grains of sand erode the soft iron oxides down to bare metal. Occasionally, plastic pellets are substituted for sand. Several large structural items from Bluff Furnace were cleaned in this manner.

Once cleaned, the artifacts must be classified. At this stage of the laboratory work, each artifact is identified in terms of its function. We knew of no well-defined blast-furnace artifact typologies generated from other sites to refer to, so we started from scratch. Of course, the artifact types defined by the archaeologist depend on what questions are being asked of the data. As an example, a typology aimed at distinguishing differences in the ages of artifacts would probably differ greatly from one that emphasized functions or styles. At Bluff Furnace there were two obvious archaeological components, both of which produced artifacts: the antebellum industrial-furnace period, which could be further subdivided into the charcoal and coke phases, and a later domestic-oriented occupation that dated primarily to the early twentieth century. Our artifact classification schemata reflect this broad division in the site's temporal and functional attributes. Nearly fourteen hundred artifacts associated with the twentieth-century domestic occupation were recovered and analyzed, but this is not the place to discuss them.[3] This is a book about a blast furnace, so we will concentrate on artifacts that are believed to be associated with the construction, operation, and demise of the industrial component.

Except in a few rare cases, such as in recovering portions of the salamander, the only way to tell whether an artifact was definitely associated with the charcoal or coke phase of the site's history was through its association with an obvious depositional event; the industrial artifacts did not lend themselves to fine temporal distinctions independent of context. However, the contextual information accompanying an artifact was often ambiguous or insufficient to precisely establish the phase to which it belonged. Since all artifacts occurring beneath the demolition fill at the site could be confidently dated to 1854–64, there was no compelling need to construct artifact typologies that would be sensitive to temporal factors.

Table 5.1

Bluff Furnace Artifact Groups and Classes

Category	Furnace Area	Casting Shed	Steam-Boiler Area	Charging Deck	Total
Wrought-iron tools	5	0	2	2	9
Wrought-iron fasteners					
Nails—wrought	15	0	0	2	17
Nails—machine cut	160	11	59	66	296
Staples	2	0	0	1	3
Spikes	4	0	1	3	8
Hinge strap	1	0	0	0	1
Bolts	27	1	6	15	49
Anchor bolts	4	0	0	1	5
Nuts	3	0	0	19	22
Washers	1	0	0	0	1
Pegs	3	0	0	1	4
Rivets	32	0	0	0	32
Totals	252	12	66	108	438
Cast-iron fasteners					
Washers	3	0	1	14	18
Hinges	0	0	0	2	2
Pulley hook	1	0	0	0	1
Totals	4	0	1	16	21
Miscellaneous wrought iron					
Masonry anchor	1	0	0	0	1
Clamps	1	0	0	0	1
Wedges	1	0	0	0	1
Rods	15	0	0	1	16
Bands	2	0	0	8	10
Bars	2	0	0	0	2
Chain	1	0	0	0	1
Rings	1	0	0	0	1
Wire	1	0	0	0	1
Unidentified	3	2	0	0	5
Totals	28	2	0	9	39

Table 5.1, continued

Miscellaneous Cast Iron					
Bracket	0	0	0	1	1
Pipe segments	30	0	0	0	30
Bars	6	0	0	0	6
Slot keys (?)	3	0	0	0	3
Structural, unidentified	102	1	1	0	104
Flat, unidentified	90	0	0	22	112
Scrap iron	46	0	13	7	66
Totals	278	1	14	30	323
Cast pigs	4	0	0	0	4
Waste materials (frequency)					
Splash iron	64	0	11	99	174
Channel iron	10	0	0	0	10
Salamander	7	0	0	0	7
Totals	81	0	11	99	191
Waste materials (weight in grams)					
Cinder, fused material	1,054	0	0	0	1,054
Slag (green)	1,182	0	0	0	1,182
Slag (black)	19,203	0	0	228	19,431
Conglomerate	14,556	0	0	0	14,556
Totals	35,995	0	0	228	36,223
Soil materials (weight in grams)					
Casting sand	4,259	0	0	0	4,259
Lime	2,843	0	0	0	2,843
Other soils	15,384	0	0	0	15,384
Totals	22,486	0	0	0	22,486
Raw materials (weight in grams)					
Charcoal	30	0	0	0	30
Coke	11,972	0	0	15	11,987
Coal	1,706	0	24	96	1,826
Iron ore	1,154	0	389	0	1,543
Totals	14,862	0	413	111	15,386

Table 5.1, continued

Other artifacts					
Ceramics	98	0	0	0	98
Glass	14	0	0	8	22
Kettle, iron	7	0	0	0	7
Personal items	3	0	1	0	4
Arms	1	0	0	0	1
Window glass	0	0	2	29	31
Wood	18	0	3	0	21
Rubber	3	0	0	0	3
Lead scrap	1	0	0	0	1
Totals	145	0	6	37	188
Bone (weight in grams)	910	0	0	1	911

More useful would be a classification format that emphasized the functions the artifacts served. This is what is presented in table 5.1, which lists the nine groups and fifty-two classes of artifacts defined for the industrial component at the site.[4] Two of the groups, wrought-iron tools and cast pigs, contain only a single class each. No attempt was made to further subdivide types within the defined classes since the functional associations of the artifacts were adequately reflected at the group and class levels.

Wrought-Iron Tools

Most of the identifiable tools recovered from Bluff Furnace are illustrated in figure 5.1. This wrought-iron artifact group-class includes, from the charging deck, a heavily worn straight peen hammer head and half of a pair of round-bit tongs. (The grasping end of the tongs is curved to receive a cylindrical shape, such as the end of a bar.) A word of caution is appropriate here: while the presence of these two tools is not inconsistent with the industrial function of the site, the tools were found in the shallow deposit over the charging-deck area and could conceivably have made their way into the archaeological record after the demise of Bluff Fur-

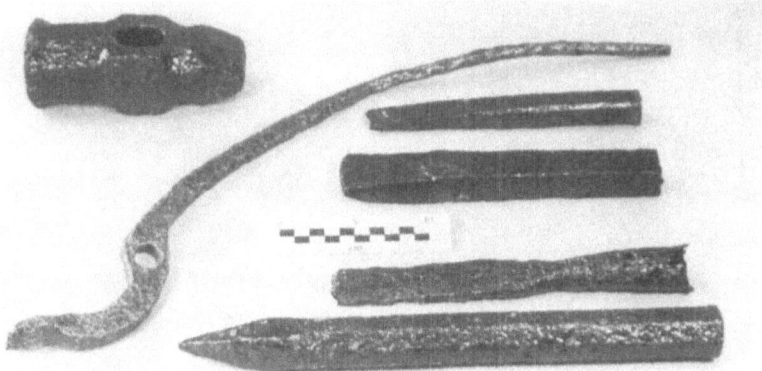

Fig. 5.1. Wrought iron tools. From upper left to lower right are: a straight peen hammer head, a round-bit tong, a square-sectioned chisel, a broken chisel shaft, a framing chisel, and an octagonal-shafted chisel.

nace. The remaining tools were found in better-controlled contexts. These include an octagonal-shafted spike or "point" chisel, showing virtually no use wear; a square-shafted chisel; a chisel-bladed tool with a socket fitting (probably a framing chisel);[5] and a broken chisel or shim with a blunted end. Except for the square-shafted chisel, which was found in the boiler area, all were recovered from the vicinity of the furnace. Not illustrated were a rat-tailed file and the handle fragments for two unidentified tools.

As this brief list indicates, few tools were present at Bluff Furnace. A low number of furnace-related tools is exactly what might be expected if the furnace was slowly and systematically abandoned. Useful items, including hand tools, would be retained by their owners, with only lost or purposefully discarded (i.e., broken) tools ending up in the archaeological record, and this seems to be the case at Bluff Furnace.

Iron Fasteners

The iron-fasteners group contains a wide variety of identifiable artifact classes, all of which are related to the joining together of wood, metal, or stone structures. Included under this category are obvious fasteners such as cut and wrought nails, staples, spikes, rivets, pegs, etc. The distribution of nails throughout the site is instructive. The relative dearth of nails from the casting shed indicates that this area probably contained few

Fig. 5.2. Miscellaneous fasteners. From top to bottom are, left: a spike-shanked door pintle, a pulley hook, and an unidentified cast-iron hook; right; a bolt, washers, and nuts.

wooden structures compared to the other areas. However, unequal samples were derived from the four excavated sections of the site: the area excavated in the casting shed was much smaller compared to the furnace-area sample. How can we tell if the low nail frequency for the casting shed is a result of actual furnace-period behavior or simply decisions by the archaeologists on where and how much to dig? One simple way is to divide the number of nails by the total area excavated in each section of the site. This produces nail frequencies for each area that are directly comparable because the figure obtained is a ratio for the number of nails *per square foot* of area excavated. Calculating the nail ratios produced the following results: the charging deck had .90 nails per square foot; the steam-boiler area, .35 per square foot; the furnace area, .21 per square foot; and the casting shed, .04 per square foot. Thus, the absence of nails in the casting shed does not appear to be contingent upon small sample size.

The charging deck, with numerous wood structures apparent in the photograph and the *Harper's* illustration, has nearly three times as many nails as the next most frequent area. This analysis also reveals a higher nail ratio in the vicinity of the furnace than what might be predicted based on our knowledge of furnace construction and operation. Given that dry wood in close proximity to the furnace would present a definite fire hazard, an emphasis on structural metal and masonry elements, rather than nailed wood, is certainly a reasonable expectation. While the pres-

Fig. 5.3. Fasteners and fittings. Top: a wrought-iron bolt with threaded ends, square nut, and heavy cast iron washer. Middle: a jointed section of wrought-iron pipe 1.5 inches in diameter. The expanded flange was filled with lead, sealing the joint, which was very likely under pressure. Bottom: a wrought-iron masonry anchor with a bolt or pintle slot. This fitting came from the stone apron near the mouth of the furnace, where it served as a secure mount for some type of equipment fitted into the round slot. All artifacts are from the furnace area.

ence of at least one (water-filled) wood structure adjacent to the furnace has already been verified, it should also be pointed out that nails are small and easily transported, and that redeposition of these small artifacts through the combination of erosion and gravity was possible from the charging deck above.

Some other common artifacts occurring under the fastener category are the wrought- and cast-iron nuts, bolts, and washers illustrated in figure 5.2. Not surprisingly, the majority of these were recovered from the furnace and charging-deck areas, illustrating once again a degree of correspondence between artifacts and the behavior generating them that is not always found at other types of sites. Not so obvious but still qualifying under the fastener category are the wrought-iron pulley hook[6] and the door pintle[7] shown in figure 5.2, as well as the enigmatic cast-iron hooklike piece that we are unable to identify precisely. The large bolt, nut, and washer illustrated in figure 5.3 are also technically fasteners.

142 / Industry and Technology in Antebellum Tennessee

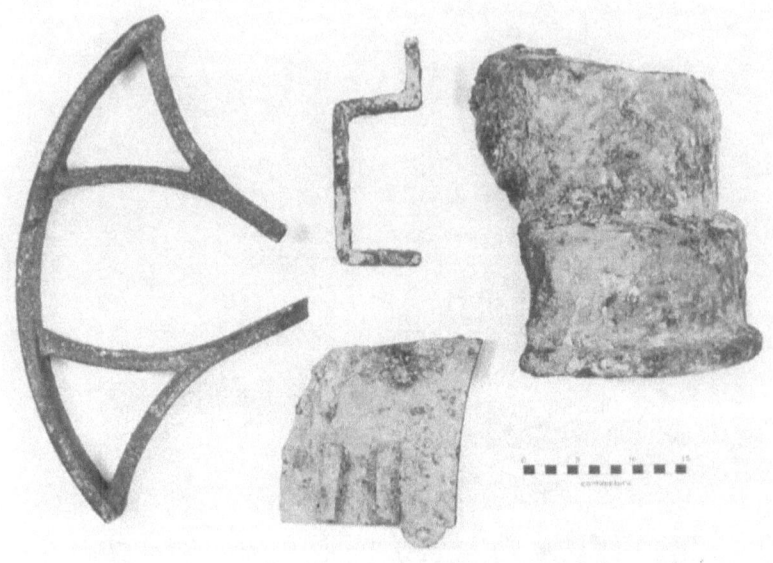

Fig. 5.4. Machine parts and fittings. Left: a light cast-iron flywheel fragment. Top center: an angle bracket or beam stirrup, on edge. Bottom center: a portion of a machine frame. Right: a large-diameter cast-iron pipe joint fragment, possibly from the blast pipe. All artifacts were found in the furnace area of the site.

Miscellaneous Iron

As seen in table 5.1, the miscellaneous group contains a wide assortment of wrought- and cast-iron classes, with 85 percent of the artifacts associated with the furnace and charging-deck areas. Of particular interest to us were the cast-iron items. Many of these objects were fragmentary and, as a consequence, unidentifiable as to specific function, but they are assumed to have been part of the structure of the furnace itself, associated structures, or machinery housing or mounts. Due to poor preservation, sixty-six items could not be identified at all, and as a consequence were relegated to the "scrap iron" class.

The most recognizable class was made up of the thirty pieces of cast-iron pipe, most with an inner diameter of one inch, that were recovered in the furnace area and are believed to have carried water to and from the furnace tuyeres. Included in this total are the pipes that connected the

Fig. 5.5. Structural cast iron. These three pieces of cast iron were fragments from long structural beams used as part of some furnace equipment or as architectural elements in the furnace buildings.

hearth to the wooden box. Two joined pieces, one with a receiving flange, are illustrated in figure 5.3. Several presumed machine parts and fittings were identified, including the flywheel and frame fragments and the large-diameter pipe joint—possibly part of a blast pipe—shown in figure 5.4. The three cast-iron pieces in figure 5.5 probably represent a class of structural iron equivalent to girders of joists, while the castings in figure 5.6 could be interpreted as evidence of direct casting over and above pig-iron production at Bluff Furnace. The flawed example on the left was found in a charcoal-period context and is quite similar to the fragment on the right, which was discovered adjacent to the exterior casting shed wall. Without additional confirming documentary or archaeological evidence, however, these two artifacts provide scant justification for a direct casting claim. Certainly none of the gates, risers, or sprues that might be expected if direct casting was going on were present in our sample. Based on the artifact analysis, the only direct evidence of casting is that associated with the production of pig iron.

The "miscellaneous wrought iron" group contained an assortment of pieces, such as wedges, rods, rings, etc., that could have functioned in a variety of ways. The single exception to this is the wishbone-shaped masonry anchor in figure 5.3, which was found embedded in the charcoal-furnace stone foundation. Its function as a structural brace was unambiguous.

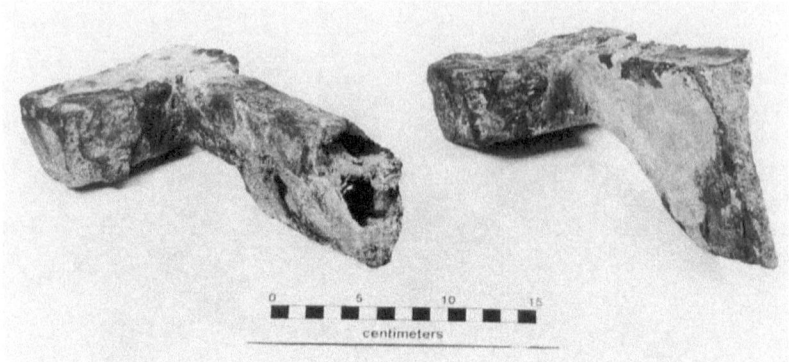

Fig. 5.6. Two castings. Both fragments may be described as rib or girder fragments; that is, they were structural or architectural in function. The example at left has a bubble in it, a serious casting flaw. This may be an indication that direct casting was taking place at the furnace, although no gate or riser fragments were found among the excavated scrap metal.

Pig Iron and Waste Materials

Four cast pig-iron bars, representing the "finished product" of Bluff Furnace, were recovered from the furnace area (three are represented in figure 5.7). One was associated with the charcoal period, one with the coke period, and the temporal affiliation of the other two was impossible to establish precisely. Two of the pigs were subjected to chemical analysis, as outlined below. Also found were several fragments of channel iron, formed when molten iron in the sand channels leading to the pig beds set up, and splash iron, formed when molten iron spilled and then cooled and solidified outside the channels and beds. Interestingly enough, none of the latter material was found in the casting shed while the majority of the collected fragments was from the charging deck. This suggests that splash iron was recycled into the charge. Another possibility is that it was produced during the construction of the charging deck. Specifically, the twenty-nine anchor bolts described in the last chapter were all set in oversized holes, at the top of the bluff, that were then filled with molten iron. The splash iron we recovered could have been left over from this anchoring process. In any event, together with the pig iron, the channel and splash iron provide conclusive evidence about the primary function and ultimate product of the Bluff Furnace operation.

Fig. 5.7. Pig-iron bars. The largest bar, at top, was found under the working floor of the coke furnace, meaning that it was probably produced during the charcoal period of operations.

Large amounts of green and black slag, two examples of which are shown in figure 5.8, were recovered from the furnace area for special analyses (see below). Although cinder and coal clinker ("fused material" in table 5.1) would logically be expected to occur in abundance in the vicinity of the steam boiler, none was found there. As a matter of fact, the "ash basins" identified in that area were devoid of any combustion by-products, including charcoal ash. Since regular maintenance of the steam boiler would have been an absolute necessity, we can suggest only that the basins were regularly cleaned and that this accounts for the absence of *in situ* clinker and cinder remains. This finding also presents a caution against assuming that one-to-one artifact and behavior correlations are always possible at archaeological sites.

Raw Materials and Soils

Substances that are assumed to be raw materials used in the iron-smelting furnace or steam-generating equipment were sampled. These include charcoal, coke, coal, and iron ore; the analysis of the last three is discussed in the "Chemical Analyses" section below. The presence of iron ore in the steam-boiler area indicates that stockpiling of ore occurred near there.

Fig. 5.8. Green slag. Slag is probably the most diagnostic of all the furnace samples; the chemical makeup of this smelting by-product is a direct indication of the operations of the blast furnace. The example at left solidified in a narrow channel, perhaps the casting channel from the hearth to the casting shed.

An area of light brown fine casting sand adjacent to the forehearth was sampled. Consisting of a channel leading from the hearth to the casting shed (refer to fig. 4.15), it was underlaid by a layer of black sand (used casting sand?) containing charcoal and was bordered by grey and brown sand containing charcoal and splash-iron fragments. Also sampled was the furnace "working floor" to the east and south of the hearth. This surface was found to consist of black sand with coke, coal, charcoal, slag, and iron fragments (furnaces are great consumers of their own by-products). West of the forehearth in the casting shed was a mass of soft lime, and this too was sampled for special analysis, as summarized below.

Other Artifacts

Other artifacts include those normally associated with domestic sites, such as bottle glass and ceramic fragments, window glass, sawed and cut bone, etc., but that were found in association with the industrial antebellum remains. Almost all are from the furnace area. The personal-items class consists of buttons, a horn comb, and a presumably intrusive 1913 penny found in the demolition fill around the hearth. One of the buttons is made of brass and appears to be of a military type: on one side is an eagle with outstretched wings. The seven iron-kettle fragments again suggest at least the possibility of direct box casting. On the other hand, they were recovered from the furnace area, not the casting shed, and they all appear to be from the same vessel. Clearly associated with the industrial function of the site were the wood fragments, the majority of which came from the wooden box and associated drain next to the furnace.

The appearance of domestic materials in the artifact inventory of Bluff Furnace can be attributed to (1) the presence of an earlier domestic occupation; (2) furnace laborers; (3) intrusion from a later occupation; or (4) some combination of these factors. Except for the obviously out-of-place twentieth-century penny, most of the artifacts in this category were probably present at the site prior to its demise. A combination of the first and second possibilities can therefore best explain the domestic assemblage. Personal items and food bone could conceivably have been generated by the furnace workers; window glass and ceramics are more likely to have been associated with a nearby dwelling.

The Chemical Analyses

Numerous samples of iron, slag, coke, coal, lime, and other substances were collected in the field for detailed chemical analyses. Three local companies, the Chattanooga Coke and Chemical Company, the United States Pipe and Foundry Company, and the American Cast Iron Pipe Company, were gracious enough to perform the needed tests on the Bluff Furnace samples. Most of the chemical analyses were aimed at constituent determination: what elements or compounds are present and in what percentages. Pig-iron samples were also examined for their physical structure through the process of photomicrography and acid etching. Some general information on these types of analyses is useful.

In *titrimetric analysis*, a small sample of the material to be analyzed is dissolved in acid and reacted with known volume of an appropriate chemical called a *titrant*. The chemical reaction between the titrant and the element being analyzed for produces a distinctive color; the quantity of the element being analyzed is determined from the amount of titrant used to obtain this color. In *spectrographic analysis*, a small amount of the sample is processed into aerosol form and then passed through a flame. Every element when it is burned gives off a distinctive frequency, or color, of light or spectrum. The range of colors or light spectra given off by the sample is analyzed and particular elements identified by their characteristic spectra.

When molten iron solidifies, the types of elements and compounds within the iron interact and form a type of crystalline structure. The amount, shape, and distribution of microconstituents, especially carbon, present in iron affect its working properties. These characteristics can be seen using *photomicrography*. To prepare iron for photomicrography, a very small sample is cut to yield a quarter-inch to half-inch area that can be highly polished for viewing under a microscope. The matrix microstructure may be enhanced by acid etching, that is, submerging the sample in a chemical such as Nital, a solution of nitric acid and ethyl alcohol.

Photomicrographs are taken with a camera mounted on top of the microscope. In general, slow cooling of cast iron produces metal containing ferrite; the iron is low in strength but hard. Iron carbide or cementite is a particular combination of iron and carbon which, if predominant in the structure of the iron, causes the iron to be hard and brittle, with high wear resistance. It indicates that the iron cooled rapidly or had a high carbon or silica content. Rapid cooling may also produce a *pearlitic* form of carbon in the iron matrix. This is seen in a photomicrograph as an aggregate of thin platelets of ferrite and cementite. Pearlitic iron is moderately strong and hard. Graphite carbon in iron comes in various forms with the flake form most common in cast iron. Flake graphite causes planes of weakness in the iron, thus requiring further heat processing of the metal to produce spheroidal graphite.

Pig-Iron Samples

Two pig-iron samples were subjected to chemical and physical analyses: one small unprovenienced pig and the larger pig from under the floor around the furnace base. Both were analyzed for silica, manganese, phosphorus, sulfur, total carbon, and iron. The crystalline matrix of each was examined through photomicrography.

The chemical profiles of both examples were well within the limits given by Johnson for a good foundry-grade iron; a carbon content of near 4 percent with a low sulfur content indicates a strong finished product.[8] Next to carbon, silica is the most important component of cast iron. The

Table 5.2

Analysis of Bluff Furnace Pig-Iron Samples

Specimen/Provenience	Constituent (Percentage by Weight)					
	FeO	Total Carbon	S	P	Mn	Si
Small pig (1-1)[a]	92.70	3.81	0.013	1.25	1.60	0.59
Large pig (1-C-44)[b]	93.50	3.94	0.008	0.83	0.30	1.34

[a] Matrix: 85% pearlite, 15% steadite.

[b] Matrix: 88% pearlite, 10% steadite, 2% ferrite.

Analytical techniques: Si, Mn, and P–ASTM titrimetric; Total Carbon and S–Leco; Matrix–photomicrography.

silica content of both pigs was well within the range given by Johnson (1 to 4 percent) to produce a good-quality gray iron.[9] This conclusion is also supported by the photomicrography of the iron, which showed a highly pearlitic matrix.

Sulfur is a troublesome element in cast iron, especially that produced with coke as a fuel; it tends to cause shrinkage and cracking of the cooled iron.[10] The sulfur content of both pig-iron samples was considerably lower than for the iron described by Overman, including that from charcoal furnaces (table 5.2). In the case of the larger pig-iron bar, this was to be expected since we assumed it had been produced during the charcoal-fired period at Bluff Furnace. The low sulfur content of the smaller pig-iron bar, presumably produced with coke, indicates that the smelting of iron in 1860 had been successful; the use of sulfurous coke had been controlled.[11]

In general, phosphorus tends to increase the fluidity of the iron and cause brittleness when the content is about 1 percent and higher.[12] Although phosphorus was relatively high in both samples, Johnson stated that "the Southern irons which carry in the neighborhood of 1 percent phosphorus command a premium for foundry purposes."[13] The phosphorus content of both pigs was certainly not outside the range reported for other nineteenth-century furnace pig irons discussed by Overman.[14]

Iron-Ore Samples

Historical data indicated that the iron ores smelted in Bluff Furnace were obtained from the dyestone or hematite outcroppings that appear along the east edge of the Cumberland Escarpment; in geological terms, the ore is part of the Rockwood Formation.[15] The ore beds at Whites Creek, near Eagle Furnace, are part of the Half Moon Island range of dyestone ridges, and most accounts indicate this as the source of iron ore for Bluff Furnace. Other accounts suggest that ore beds directly across the river from Chattanooga were used in the furnace.[16] In Lesley's 1859 compendium, the term "fossil dyestone" is used to described the furnace ore, indicating use of a calcareous variant of Rockwood ore containing aggregates of fossil organic forms and clay.[17] Variability in iron ore is common, but Rockwood ore tended to be red to bluish-black, lustrous, and mixed with calcium carbonate, silica, alumina, magnesium carbonate, and a variety of minor elements.

Table 5.3

Analysis of Bluff Furnace Iron-Ore Samples

Specimen/Provenience	Constituent (Percentage by Weight)					
	Insoluble Matter	Fe	P	Mn	SiO_2	Al_2O_3
#1 (3-B-5)	7.90	59.03	0.25	0.18	6.06	1.80
#2 (3-B-7)	9.00	58.03	0.20	0.49	6.60	2.40

Analytical technique: ASTM titrimetric.

Chemical analyses run on the iron ore samples retrieved at Bluff Furnace focused on the silica, manganese, alumina, and iron (as FeO) constituents in the samples (see table 5.3). The percentages of each fell within the anticipated values of Rockwood ores, the historically suggested source of the iron ore. The chemistry of the Bluff Furnace ores indicates that the hematite being smelted had a higher metallic content than is usually found in Rockwood ores, and higher on the average than other hematite ores smelted in Great Britain and other parts of the United States.[18] This suggests that only the highest-quality ore was mined and processed for use in the furnace. Rockwood ores were considered non-Bessemer grade, containing 0.25 to 0.75 percent phosphorus.[19]

Coal and Coke Samples

Analysis of bituminous coal samples taken from Bluff Furnace was particularly important. The 1859 Lesley compendium and a deed reference suggested that the furnace coke used at Chattanooga might have been made from local Raccoon Mountain coal. Both of these sources, however, referred to intended *future* use of the coal. Later documents confirm that coke was being produced at the Etna Mines on Raccoon Mountain but that it was foundry-grade coke and not suitable for furnace use. Still another commentary indicated that the coke actually consumed at Bluff Furnace came from the Roddy banks near Eagle Furnace in Roane County; moreover, the coal was coked *before* shipment to the bluff—meaning that raw coal for use in the furnace (as coke) should not be present at the site.

One question we hoped to address with an analysis of the coal samples from the archaeological site was where the coal/coke used at the Bluff Furnace was mined. There are several complications that arise in trying to answer this question. As with iron ore, the chemistry and physical structure of any coal varies considerably within a single coal seam, making comparisons tenuous. We also cannot exclude the possibility that raw coal from Raccoon Mountain was used at the bluff in steam generation. There is no information suggesting that any coal was coked at the furnace site; thus we have no real reason to assume that coal present on the site was intended for use as a blast fuel. Moreover, the chemical composition of the coal would be significantly altered in the transformation to coke, making the connection of coke samples with coal mine sites difficult if not impossible.

Table 5.4

Chemical Analysis of Coal Samples from Bluff Furnace and Related Mine Sites in Marion and Roane Counties

Specimen	Constituent (Percentage by Weight)[a]			
	Volatile Matter	Fixed Carbon	Ash	Sulfur
Bluff Furnace[b]	27.1	65.4	7.5	2.6
Whiteside Mines[c]				
Old Etna Bed	25.3	67.4	2.1	0.6
Old Etna Bed	26.3	66.7	2.5	0.7
Old Etna Bed	26.7	64.7	2.9	1.1
Castle Rock Bed (av.)	23.4	62.0	10.8	2.5
Kelley Bed (av.)	27.6	61.7	7.2	1.3
Average	25.7	64.5	3.1	1.2
Bon Air (maximum)[d]	39.4	67.4	23.8	5.1
Bon Air (average)	33.1	54.5	9.8	2.4
Bon Air (minimum)	24.6	46.4	2.1	0.4
Sewanee (maximum)[d]	33.1	64.1	23.1	2.0
Sewanee (average)	29.0	57.0	10.8	0.7
Sewanee (minimum)	22.9	47.2	6.5	0.5

[a] Analytical technique: ASTM titrimetric.
[b] Provenience: 1-C-34.
[c] *Source*: Nelson, *The Southern Tennessee Coal Field* (1925), 162. All samples near Whiteside, Marion County.
[d] *Source*: Luther, *The Coal Reserves of Tennessee* (1959), 201, 242. Bon Air samples from Marion County; Sewanee samples from Roane County.

Table 5.5

Chemical Analysis of Bluff Furnace Coke Samples

Sample Color	Constituents (Percentage by Weight)		
	Fixed Carbon	Ash	Sulfur
Dull black[a]	87.44	11.06	0.73
Gray[a]	80.00	18.42	1.28
Silvery gray[b]	76.52	21.18	1.24
Dull black[c]	86.87	11.73	1.23
Silvery gray[d]	88.72	9.80	0.98
Silvery gray[d]	72.06	25.25	1.44
Average values:	81.94	16.24	1.15

[a] Provenience: 1-C-34.

[b] Provenience: 1-C-19.

[c] Provenience: 1-C-42.

[d] Provenience: 1-C-38.

Table 5.4 presents the analysis of a coal sample from Bluff Furnace, compared with coals from the Whiteside mines and Bon Air deposits on Raccoon Mountain, and the Sewanee seam from Roane County (the vicinity of Eagle Furnace). The bluff sample falls within the range of variability of Bon Air coals in all categories, and within the Sewanee range in three. Taking the samples from the vicinity of Whiteside as a whole, the Bluff Furnace sample falls within the range of variability in every case. Because only one bluff coal sample was analyzed, no firm conclusion can be reached about the origin of the coal. Only an analysis of rare trace elements in regional coals would address the question of origin in a conclusive manner, and the raw data for such an analysis are not available.

Six coke samples from Bluff Furnace were analyzed for the percentages of fixed carbon, ash, and sulfur present in the fuel (table 5.5). Ideally, good blast-furnace coke was to contain 0.8 to 1.5 percent sulfur, 85 to 90 percent fixed carbon, and 8 to 12 percent ash.[20] The fixed carbon in the bluff samples was somewhat less than ideal, at least by early twentieth-century standards, and the ash somewhat higher. Comparable data from mid–nineteenth-century cokes would have been more useful if the success of the Bluff Furnace coking process were to be assessed in relative terms.

Slag Samples

Although nominally a waste by-product of the smelting operation, slags are important indicators of the relative efficiency of the blast furnace; their composition reflects the reduction process taking place in the stack. Seven slag specimens from Bluff Furnace were analyzed for chemical content (see table 5.6). Four of the samples came from above the working floor of the coke furnace, and three were from the slag heap north of the casting shed.

As groups of samples, the slags recovered from around the coke hearth on the average differ chemically from the samples drawn from the slag heap on the riverbank. At present the reason for this difference is not clear. The riverbank slag samples may have been drawn from deposits of the charcoal-fired smelting period of the furnace. Alternatively, their different chemistry may be the product of post-depositional factors such as groundwater leaching. What is clear is that the iron content from the coke-hearth slags is higher than what was generally considered accept-

Table 5.6
Chemical Analysis of Bluff Furnace Slag Samples

Specimen	Constituent (Percentage by Weight)							Desulfurization Index $\dfrac{CaO+MgO}{SiO_2+Al_2O_3}$	Refractory Index $\dfrac{Al_2O_3}{CaO+SiO_2}$
	SiO_2	Al_2O_3	CaO	MgO	MnO	S	FeO		
1[a]	41.20	13.10	30.70	6.58	0.33	0.54	7.27	0.69	0.18
2[a]	39.60	13.40	30.10	6.60	0.33	0.56	9.18	0.69	0.19
3[b]	41.50	13.40	30.30	5.85	0.38	0.51	7.77	0.66	0.19
4[c]	34.30	10.60	25.90	4.49	0.35	0.57	23.70	0.68	0.18
Average[e]	39.20	12.60	29.20	5.88	0.35	0.54	11.98	0.68	0.18
5[d]	45.90	14.20	32.30	3.96	1.02	0.10	1.74	0.60	0.18
6[d]	43.20	14.40	36.30	4.05	0.45	0.08	0.99	0.70	0.18
7[d]	41.70	14.20	38.30	4.48	0.39	0.12	0.38	0.77	0.18
Average[e]	43.60	14.30	35.60	4.16	0.62	0.10	1.04	0.69	0.18

[a] Furnace base (1-C-34).
[b] Furnace base (1-C-36).
[c] Furnace base (1-C-38).
[d] Slag pile (1-E-7).

able in nineteenth-century practice. Fairbairn indicated that these black, iron-bearing slags are formed when the supply of coke is insufficient to deoxidize all of the iron, and a portion of the iron passes into the slag, giving it a black or very dark green color.[21]

The color, texture, and porosity of slags vary with a number of factors and their combinations, including chemical composition, temperature of formation, and rate of cooling. In general, slags with high silica contents tend to be darker and more glassy than more basic slags. Cold-blast slags and those that cool rapidly tend also to be more vitreous. An increase in porosity is often associated with a lighter color and a decrease in vitreosity.[22] Too little is known about Bluff Furnace to correlate the physical properties of slag with furnace operation. Sample characteristics discussed below are offered as a matter of general interest.

The most important role of the slag was to remove sulfur, which entered the burden as constituents in the fuel and ore. Compounds that removed sulfur were, in descending order of importance: lime, magnesia, alumina, and silica. One measure of the chemical "success" of the slag is the *desulfurization index*, computed by dividing the percentages of lime (CaO) and magnesia (MgO) in the slag by those of silica (SiO_2) and alumina (Al_2O_3).[23] In comparison to slag samples from other sites (compare table 5.6 with Appendix 3), the Bluff Furnace slags were relatively good at desulfurization, an indication that the use of sulfurous southern coal as coke was being properly managed in order to yield good pig-iron chemistry. To a point, a high desulfurization index leads to a lower slag viscosity and free-running temperature, meaning that the slag flowed easily over the damstone of the furnace.

Comparing twentieth-century data on slags with Bluff Furnace samples permits us to estimate the temperature range at which the slags flowed from the Chattanooga furnace. Bluff Furnace slags contained 12 to 15 percent alumina and 33 to 60 percent silica, indicating that the free-running temperature of the slags was in the range of 1340 to 1370 degrees Celsius (2440 to 2500 degrees Fahrenheit). By early twentieth-century standards, this range was within the low critical temperatures then considered optimum.[24] Following John White's example, we have attempted to determine the temperature of formation of the Bluff Furnace slags.[25] Using the 15-percent alumina place of the lime-magnesia-silica tetrahedron given by White and developed by Osborn and used by White and others, we find the Bluff Furnace slags falling close to the 1400-degree-Celsius isotherm. In simple language, the slags indicate that a furnace temperature of 1400 degrees Celsius (2550 degrees Fahrenheit) was reached at Bluff Furnace.

Also after White, we have computed the refractory index of the Bluff Furnace slags. The *refractory index*, computed by dividing the percentage of alumina by those of lime and silica combined, describes the refractory character, or fusibility, of the slag. The higher the refractory index, the higher the operating temperature of furnace required to keep the slag in a fluid state. The refractory indexes for the Bluff Furnace samples are consistent with those of similar period furnaces (see Appendix 3).

Lime Sample

A sample drawn from the deposit of the white, chalky substance west of the furnace base confirmed that lime was produced at Bluff Furnace during the Civil War. The sample was 55.68 percent CaO, which is calcium oxide or lime. Other constituents present were magnesia, MgO (1.14 percent), silicon, SiO_2 (12.12 percent), sulfur, S (1.58 percent), and carbon dioxide, CO_2, (5.68 percent). The sample lost 16.92 percent of its mass as bound water when ignited at 600 degrees Celsius.

Summary

Chemical analyses of iron, coke, coal, and slag samples permitted us to make some empirical statements about the smelting operation at Bluff Furnace. First, the origin of the coke used at Bluff Furnace remains a moot point. Historical documents suggest the possible use of Raccoon Mountain–Etna Mines coal for coke while other sources state that coal from the Rockwood area upriver was coked for use in Bluff Furnace. Analysis of the one coal sample from the bluff is inconclusive as to its origin, and in their altered state the bluff coke samples cannot be used to address this question.

Slags produced during the coke-fired smelting operation at the bluff suggest that, while desulfurization was being adequately managed, the reduction of the iron content from the ore was perhaps not as efficient. This less-than-optimum iron recovery rate was probably not a serious drawback given the relatively high iron content of the ores being used. Pig-iron samples indicate that, in the charcoal period of Bluff Furnace and probably during the coke period, good-quality foundry-grade pig iron was being produced at the furnace.

These simple but scientifically obtained observations demonstrate that Bluff Furnace actually functioned as it was supposed to, producing quality pig iron using coke as a fuel. Despite its limited output and short life span, the furnace did achieve a certain amount of technical success.

The laboratory analysis of ore, coal, and coke samples also defined some limitations of the method of comparative chemical analyses in determining the point of origin of these blast materials. Without baseline samples of these materials from the possible sources, the significance of the comparisons was tenuous. Samples of coal and coke from the Cravens coal concession at the Etna Mines, for example, would have been useful to address the problem of the point of origin of the coal and coke used at Bluff Furnace.

CHAPTER 6

Bluff Furnace at Chattanooga: A Retrospective

Historical and archaeological data on Bluff Furnace are two distinct lines of evidence that intertwine around a single, but now remote, series of events. The degree to which these lines of research converge reflects our closeness to that highest of abstractions, the truth. But in truth, interpretations of the past continually evolve with the changing perspectives of the viewer. As we shall see, modern theories about the nature of technology and enterprise in the antebellum South have dramatically changed in recent decades.

Historian Charles B. Dew noted a quarter of a century ago that "a thorough history of the Southern iron industry during the nineteenth century is badly needed,"[1] and there has been some effort to correct this situation. Much of the available historical literature on the southern iron industry is, however, either topical (concerning the Civil War period, for instance) or of limited geographical scope (such as state-focused studies).[2] Similarly, archaeological literature on antebellum ironworks in the South is virtually nonexistent. Intensive excavations of iron-producing sites in the region are extremely rare.[3]

Our study of Bluff Furnace at Chattanooga has added to the limited literature a detailed case history of technological innovation and change in the iron industry of antebellum Tennessee. Compared to historical accounts of the furnace operation, the "unedited" archaeological data from the furnace site has provided new insights into the operations of the South's first coke-fired, hot-blast furnace. As we shift our focus from an individual piece of archaeological or historical data to its larger associations, we begin to appreciate its fuller meaning.

Success or Failure?

The long, and unfinished, trail of research on Bluff Furnace was originally predicated by a survey looking for "significant" archaeological sites in the path of modern construction. A few lines in a nineteenth-century iron manufacturing history yielded a claim of uniqueness that preserved the site. But it is all too easy to weigh Bluff Furnace on a qualitative scale, balancing its historical superlative—the *first* coke furnace in the South—against its broader role in the iron industry of Chattanooga and the surrounding region. Evaluated objectively, the story of Bluff Furnace and the East Tennessee Iron Manufacturing Company is a mixture of successes and failures.

It is possible to outline the events on the bluff overlooking Chattanooga's Ross's Landing, but these "facts" are part of a larger fabric. Historical accounts of Bluff Furnace from the late nineteenth century noted, almost in passing, that the East Tennessee Iron Manufacturing Company had put into blast in 1860 the first coke-fired furnace in the southern Appalachian iron-producing region. A relatively conventional hot-blast steam-powered charcoal furnace, built in 1854 by Robert Cravens, was leased by a northern company and converted into a plant with the most current technology then in use in the United States. Once before, in 1848, Cravens had attempted this conversion, only to be met with failure and a lawsuit. Two northern ironmasters, James Henderson and Giles Edwards, may have undertaken the construction of the new furnace plant for their own corporate purposes, but all the resources of the East Tennessee Iron Manufacturing Company were marshaled to bring the coke-iron plant into blast early in 1860. With iron ore and possibly coke being barged eighty-one river miles from Eagle Furnace, maintaining stockpiles at the plant was difficult. Like the charcoal iron produced before it, the pig iron cast at the new facility was of good foundry grade—the potential chemical problems of using sulfurous, southern coals for coke fuel having no ill effect on the iron smelted. The furnace functioned satisfactorily until a blast failure occurred in November of that year. Either by chance or by neglect, the hearth inwall ruptured, and the semi-molten burden in the hearth congealed into an intractable mass. Only extensive repairs to the hearth could bring the stack back on line.

The political turmoil in the months prior to the Civil War apparently disrupted the working relationship between the northern managers and technicians and their southern counterparts. James Henderson and other skilled workers departed Chattanooga, leaving the plant to languish without their expertise. Before a restart of the furnace, military events impinged: Tennessee was invaded by Union armies. The movable furnace equipment was dismantled and reinstalled at Oxford Furnace in Alabama, leaving only the standing architecture at Bluff Furnace, including its disabled cupola-style stack. The East Tennessee Iron Manufacturing Company dissolved as a working entity in 1863, many of its principals by this date being dead or scattered before approaching armies. A deposit of lime found near the furnace confirmed the historical account that the nearly demolished stack was used for lime making during the Federal occupation of Chattanooga in 1864–65. During construction of a military bridge over the river, Federal engineers scavenged the furnace site for materials. By the end of the century, Bluff Furnace was buried and forgotten.

An understanding of the significance of Bluff Furnace goes beyond the history-making coke-fired blasts of May and November 1860. The list of facts and interpretive hypotheses presented above offers only a tiny, unframed picture with little background and very little contrast. In ever-widening circles, we should shift the focus away from Bluff Furnace alone and encompass its industrial milieu.

In the history of the southern Appalachian iron region, Bluff Furnace attained significance as the first of what would later be considered modern-style coke furnaces. Beyond this nominal significance, how do we objectively evaluate the overall economic success of the company or, more to the point, the *two* companies involved: the charcoal furnace operation of Cravens and Whiteside and the leased coke furnace plant of Henderson and Edwards?

The prime motivation for establishing a blast furnace and foundry at Chattanooga had been purely commercial; it was perceived by the investors of the East Tennessee Iron Manufacturing Company as a way to make money. The railroads at Ross's Landing led to distant, much-coveted markets, from the Mississippi River ports to the Atlantic Coast. Chattanooga's river and rail connections facilitated exploitation of the surrounding region's considerable mineral wealth. In addition to being a tool to facilitate mining and manufacturing, the railroads in themselves were important markets for furnace and foundry. The company expansion from its base at Eagle Furnace in Roane County to Bluff Furnace at Chattanooga was a literal step toward larger markets.

The choice of steam power for the Bluff Furnace, although the natural result of the lack of usable water power at the site, provided greater reliability in the powering of the blast machinery. Use of the hot blast was not particularly innovative from a regional or state perspective, either in 1854 or 1860, but it does indicate that the company was not bound by the tried and proven cold-blast charcoal tradition that obtained in other areas.[4] Nor was the ironmaster of the East Tennessee Iron Manufacturing Company hidebound by tradition; Robert Cravens had already attempted coke-iron production at Eagle Furnace.

Perhaps Cravens and his associates lacked expertise, but they did not lack vision; northern experts arrived in Chattanooga to convert the charcoal, hot-blast furnace to burn a fuel derived from locally abundant coal. The use of coke fuel represented a commitment to the newest available technology. An ironclad, cupola-style furnace was still somewhat experimental in 1860, and its construction entailed some daring on the part of the furnace operators. The thin-walled cupola stack, recycling its own waste heat to preheat the blast, could nevertheless burn hotter and smelt iron ore more cost-effectively.

Technologies notwithstanding, economics played the pivotal role in American iron production. The most successful iron and steel producers of the late 1800s increasingly practiced economies of scale, enlarging smelting plants and maximizing their outputs by mechanization and blast management techniques referred to as "hard-driving."[5] Under the regime of new technology and plant operation, the unit cost of pig iron decreased relative to the cost of raw materials. Was the charcoal Bluff Furnace larger than its contemporaries, and was it being driven harder?

By virtue of its size and hot-blast apparatus, Bluff Furnace started out as a distinctive industrial entity; even during its charcoal configuration, the plant was unusual. Using Lesley's 1859 data, Bruce Seely estimates the median height of charcoal furnaces built in the United States after 1850 to be a little over thirty-five feet, while Bluff Furnace was forty feet high.[6] The use of a hot blast also varied from the norm, for of the 711 charcoal furnaces listed by Lesley, only 271 (38 percent) used either warm or hot blasts.[7] In Tennessee, the Bluff Furnace was one of four in the Dyestone region (out of eleven furnaces in all) and one of nine in the state (from a total of sixty-two) using the hot blast. The data collected in Lesley's 1859 *Guide* indicate that in East Tennessee, the charcoal-fired Bluff Furnace had the largest stack, measuring in at 10.5 feet at the boshes and 40 feet in height. This stack was much larger than the company's earlier Eagle Furnace in Roane County.

Furnace capacity is not only an expression of the physical size of the stack but of the degree to which the smelting is driven by factors such as blast pressure and hot-blast temperatures. The only available production rate for the bluff, however, was a very modest 13.2 tons per week. Three smaller contemporary furnaces in the Dyestone region had higher pig-iron production rates. By contrast, of the Western Highland Rim furnaces in operation in the late 1850s, one was larger in stack height (10.5 by 42 feet), and all were producing pig iron at rates ranging from 23.4 to 50 tons per week.[8] What Bluff Furnace produced in 1857–58 is undocumented, leaving a considerable gap in our knowledge about production rates during the latter half of the charcoal-blast period. The large size of the charcoal stack may well have reflected a desire by Cravens and company to participate in higher furnace capacities and productivity, but the only documented production rate available suggests a mediocre performance.

It was in the conversion to coke as a fuel that Bluff Furnace was most distinct from other smelting operations in the region and nation: Swank reports just twenty-four furnaces in Pennsylvania and Maryland that were actively using coke or adapting to its use by 1856, and cupola-type furnaces were virtually unknown in this country at this date.[9] In Great Britain and in the northeastern United States, the adoption of coke was, in part, a response to the scarcity and thus higher cost of charcoal. Traditional charcoal furnaces typically consumed the forests immediately surrounding them and were then forced to haul charcoal over increasing distances at increasing cost. Was there a charcoal supply problem in Chattanooga prior to 1858? Unlike most furnace operations in Tennessee, the Bluff Furnace was sited in the midst of an urban setting, although the town at Ross's Landing was little more than a village. The residential and industrial consumption of wood in the immediate vicinity of Chattanooga may have been a factor compelling the conversion to coke, but the availability of a better fuel—coal as coke—may have been more important in the decision to change fuels.

The use of sturdy coke permitted a substantial increase in the capacity of blast furnaces. Coke furnaces could thus readily participate in the economy of scale.[10] Was the coke-fired Bluff Furnace of substantially greater capacity than its charcoal predecessor? For the coke period at the bluff, we have no data comparable to that published in the Lesley compendium, either historical or archaeological. The charcoal stack of Bluff Furnace was forty feet high, according to Lesley. On the basis of photographic comparisons between the 1858 woodcut of the charcoal furnace plant (fig. 3.5) and the 1860 stereoscope of converted Bluff Furnace (fig. 3.7), the coke stack was perhaps forty-five feet high: slightly higher than the stone charcoal stack. The coke stack was said to be eleven feet wide at the boshes,[11] making the cupola furnace wider and higher than its charcoal predecessor.

How hard was the coke furnace driven in comparison to the charcoal stack? We have no production rate (pig-iron quantity versus time) from the coke furnace, only the information that five hundred tons were produced in its first blast. Since there were no other coke furnaces in the southern Appalachian iron-smelting region, there are no plants with which to compare the known dimensions and capacities of Bluff Furnace. The relative scale of production with coke is thus unknown.

Are there other indicators of upscaled activity at Bluff Furnace, during either plant period? During its effective corporate history (1847–63), the East Tennessee Iron Manufacturing Company increased its maximum capitalization from $250,000 in 1847 to $1,000,000 in 1852. Nominally, the increase in capitalization might indicate an increase in the overall scale of the enterprise. This documented recapitalization occurred two years before the completion of Bluff Furnace, four years before the plant was put into operation, and six years before the coke conversion of the stack at Chattanooga was accomplished. Cravens, Whiteside, and company did scale up the operations of the company in order to build Bluff Furnace. Corporate capitalization of the lessees of the furnace in 1859 is not documented, however. Largely due to this lack of data, no firm connection between the level of capitalization and the use of coke can be established.[12]

Another marked trend in the economic structure of successful iron manufacturing enterprises in the nineteenth century was the increasing degree of vertical integration, the linking under one corporate structure of materials acquisition, transportation, and manufacturing. There is no evidence that the East Tennessee Iron Manufacturing Company formally vertically integrated with mining, transportation, manufacturing, and marketing interests during the course of its existence; the corporate charter nominally limited the scope of the company to the manufacture of iron. On the other hand, the makeup of the company's board of directors has led to speculation that vertical integration was effectively achieved, if not formally defined, by the principals involved.[13]

Each individual stockholder of the East Tennessee Iron Manufacturing Company brought into the concern assets from various sectors of the economy. Several individuals brought in valuable real estate, including industrial properties. Robert Cravens vested the company with title to the working Eagle Furnace and Forge, with mineral lands, in Roane County. Cravens, Boyce, and Whiteside owned coal lands on Raccoon Mountain, and Cravens held a mining concession to produce coal and/or coke for smelting and related industries in Chattanooga. William Williams had considerable investments in steamboating, and Whiteside and Cravens (and perhaps more of the investors) held stock in railroads with Chattanooga terminals. The company ran its own steamboat and barge line as a necessary adjunct to the smelting operation. The furnace lessees in 1859 obtained permission to run an industrial railway from the plant site to the main wharf, indicating their intention to improve the transportation of their products and raw materials. Some of the indicators of successful vertical or forward integration, however, seem to be negative.

Several signs point to indifferent corporate performance during the period of charcoal smelting. The company leased, and finally sold, its foundry in 1858, this despite the fact that a foundry was one of the principal consumers of the company's main product, pig iron. Sale of the foundry was vertical disintegration. The company built the Bluff Furnace at Chattanooga and yet did not "blow in" for two years, perhaps being stalled by provision and transport of raw materials. From the standpoint of the furnace, this backward integration was essential and a customary aspect of early iron plantations as well as modern integrated facilities.

Forward or vertically integrated iron facilities incorporated rolling mills to convert pig iron into finished products such as "merchant" bar iron and rail. Several of the principals clearly contemplated erection of a rolling mill at Chattanooga, evidently to be situated on the Chattanooga Creek property bought in 1852. Mention was made in Cravens's coal reserve lease at the Etna Mines of a possible rolling mill to be built in Chattanooga. There was a rolling mill built on that tract, but not by Cravens and company, who sold the railroad-frontage parcel during the Civil War. Given that Bluff Furnace was leased to James Henderson, how profitable could the charcoal furnace have been in the first place? If the charcoal furnace was losing money, the shift to coke may have been an effort to put the Bluff Furnace smelting operation "in the black" by an increase in scale.

A key concern in any risky industrial enterprise such as iron smelting is effective management. In the critical period of the late 1850s, James A. Whiteside, the company president, was in Nashville serving as vice-president and operating manager of the N&C Railroad.[14] Cravens had built one successful furnace complex at White's Creek, but the indications are that Bluff Furnace was lackluster in performance. When it came to the coke conversion, Cravens's experience in 1848 had been all bad; in 1858–60, this experience would have had little utility. But what of the Roane County holdings and the output of the Eagle Furnace and Forge? Were these plants producing a steady profit for the company? We know only that the company steamboat and barges made frequent downriver trips, hauling pig iron to markets at Mississippi ports. The absence of more detailed information is frustrating.

In sum, the investors in the original company had diverse and interlocking industrial and commercial interests, but these interests were not consolidated under one corporate structure that would have given the company absolute control over its own raw materials supply, transportation, and manufacturing of finished cast and wrought-iron products. In the absence of detailed corporate records, we must conclude that the East Tennessee Iron Manufacturing Company remained small in scale although strongly committed to using the best available technology to smelt iron ore. The original investors, including Cravens, Whiteside, and others, did not innovate in the construction of the charcoal furnace but built a large facility that incorporated the best proven technology of the day: steam power and the hot blast. The northern lessees of the furnace, represented by James Henderson, created a plant on the cutting edge of blast technology, adding the cupola stack and converting to use of the more efficient coke fuel. The smelting of iron with coke at Bluff Furnace never got beyond the experimental or trial stage; history intervened before the long-term profitability of the new enterprise could be realized.

The case history of Bluff Furnace demonstrates the interplay of technology and society; technological innovation does not take place in a vacuum but in the midst of interconnected social, political, and economic forces. In 1860, the most current blast-furnace technology had been imported to Chattanooga and made to function. But volatile regional politics disrupted the operation and resulted in dissolution of the enterprise; the *human* factor in the industrial equation was uncontrollable.

The example of Bluff Furnace is germane to the study of technological innovation in the antebellum South and the politics of the slave-based economy. We have made the point that structural failure probably terminated the final blast of the coke furnace and that James Henderson, the main individual responsible for the conversion, departed the scene for a more favorable political climate. Yet, a telling fact about the state of affairs in Chattanooga was that, despite an instant market for iron in all forms created by the ordnance demands of the Confederacy, Bluff Furnace evidently could not be brought back into production between November 1860 and the Federal occupation of the town in September 1863. In the midst of extraordinary efforts throughout the South to procure vital materials like iron, the smelting facility at Chattanooga was simply unworkable. Without the expertise to operate the coke furnace, the plant was useless. The subsequent military events in Tennessee only partially explain why the furnace was never rebuilt.

The economic effect of the Civil War on the economy of the South was disastrous, and many of the South's iron-production centers had been the special targets of Federal raiders. However, Southern pig-iron production had been flagging in the years prior to the conflict. The economic depression in 1857–58 ended a decade of sustained growth for iron production in the United States.[15] In Tennessee, a record seven blast furnaces were constructed in 1854, and four old furnaces were abandoned in the same year.[16] After the depression, Southern iron manufacturing showed little resilience in the face of withering competition with Northern iron producers.[17] Pig-iron production statistics of 1860 demonstrated the declining participation of Tennessee in the national iron industry, and with some fluctuations, this trend continued.

Chattanooga's Postwar Iron Industry

The Bluff Furnace and the company foundry were the first heavy industrial plants in Chattanooga, and they marked the beginning of the town's long association with metal production and fabrication. As the beginning of a trajectory in industrial development, the uneven success of Bluff Furnace unfortunately characterized the entire historical trend of the metals sector of the regional economy. Chattanooga's promise of becoming a major iron production center in the late nineteenth century was a phenomenon that was envisioned to a much higher degree than it was actually realized. Nonetheless, as a metal manufacturing and fabricating center, the city did acquire a significant regional if not national reputation.

Chattanooga's transportation facilities were described as superior, the bituminous coals of the region, excellent, and the iron ore, good.[18] The number of foundries and fabricating plants in Chattanooga using iron and steel multiplied rapidly during the late nineteenth century, and the modern historical stereotype of Chattanooga as a "railroad town" belies its nineteenth-century stature as an "iron town."[19] Yet, despite many favorable conditions, the iron smelting of the Chattanooga region could not keep pace either with Northern expansion and trade competition or with the later iron and steel industries of the Birmingham region. Local furnaces simply could not match their scale or operational economy.

Chattanooga's second blast furnace was built on the banks of the Tennessee River on the west side of town in 1874. The new furnace of the Chattanooga Iron Company was of modern shell design with a sixty-foot stack measuring thirteen feet, four inches, at the boshes.[20] Located on a flat floodplain of the Tennessee River, the furnace was loaded by mechanical skip-hoists. The furnace was of modest size, capable of producing twenty-five or thirty tons of iron per day using coke from Dade County, Georgia, and a mixture of regional ores. Most of the furnace output went to a nail mill in Terre Haute, Indiana; the remainder was shipped to markets in Cincinnati and Louisville.[21]

In 1885 the company was reorganized as the Chattanooga Coal and Iron Company, and construction began on a new eighty-ton furnace which produced, like most furnaces in Tennessee, a high-phosphorus, foundry-grade pig iron. Between 1895 and 1911, the furnace ran intermittently when pig-iron production was profitable. In 1911, a new two-hundred-ton furnace was built. The Chattanooga Furnace was one of the many victims of unstable pig-iron prices after World War I. In 1919 the plant closed, and in 1928 the furnace was cut up for scrap.[22]

Chattanooga's third blast furnace was erected near the Tennessee River just one mile upstream from the site of Bluff Furnace. The Citico Furnace Company was incorporated in November 1882, and the furnace was "blown in" in April 1884. The coke-fired iron-shell furnace produced foundry-grade iron from hematite and limonite ores mined from company mineral lands in northern Georgia and Alabama. Most of the furnace's product was consumed locally. The unprofitable furnace was scrapped in 1911.[23]

With the scrapping of these two blast furnaces, pig-iron production in Chattanooga ended. Considerable quantities of iron and steel products were being fabricated in the city, and foundry cupolas consumed ton after ton of pig iron, but the primary smelting of iron ore ceased. Economics, not poor technology, was at the root of this change. The experience of Chattanooga's blast furnaces was repeated at other plants in the Tennessee.

The adoption of coke in Tennessee furnaces had been slow; only in the Dyestone iron-producing region was there a rapid postwar movement toward the use of this fuel. Unlike other furnace regions in the state, the Dyestone Belt possessed significant quantities of ore, coal, and flux in close proximity—a formula for commercial success. The Rockwood furnaces of the Roane Iron Company, led by war veteran General John T. Wilder, began production of hot-blast coke iron in shell-type furnaces as early as 1867. These were the first such modern plants built in the South after the war.[24] Elsewhere, the movement toward the modern blast plants pioneered at Bluff Furnace was slow.

In 1873, only three stacks burned bituminous coal and coke in Tennessee; two of the stacks were Wilder's Roane Iron Company furnaces in Rockwood. There were five charcoal furnaces in the Eastern Iron Belt, and fourteen charcoal stacks in the Western Iron Region, not all in blast however.[25] Most of these charcoal furnaces had been built before the Civil War, and many were still using a cold blast.

In 1874, there were twenty operating furnaces in Tennessee. Five furnaces exploited the Eastern Iron Belt, which contained mainly limonite ore, and in the Dyestone Belt of hematite ores, which supplied Bluff Furnace in its day, were four furnaces; three other furnaces were reported to be under construction in 1874.[26] Limonite was exploited in Tennessee's Western Iron Belt on the Western Highland Rim of Middle Tennessee. In 1874, only eleven furnaces in that area were in operation, all burning charcoal. In 1878, most of that region's furnaces were out of blast. There were fifteen furnaces in the area said to be abandoned or idle but still capable of production.[27]

In 1876, statistics on pig-iron production demonstrated the stagnation if not decline in Tennessee's iron industry. Tennessee furnaces, twenty-four in number, produced only 24,585 net tons of pig iron, barely 1.17 percent of the national total (see table 6.1). This production level was below the state pig-iron output in 1850, a quarter century earlier. Alabama's 1876 pig-iron production was comparable to that of Tennessee (1.18 percent of the nation's total), but this yield was produced by half the number of Tennessee's furnaces (twenty-four in Tennessee, thirteen in Alabama)—clearly reflecting the economy of scale. Regional patterns of dominance in the pig-iron trade had changed little since 1850, however; in 1876, Pennsylvania alone produced 48.23 percent of the nation's pig iron.

Iron facilities other than blast furnaces also displayed troubled records in the postwar years. One of the few useful residuals of the Federal military presence in Chattanooga during the Civil War had been the rolling mill built by the United States Military Railroad in 1864–65. The plant was furnished with ironmaster John Fritz's rail mills.[28] At this facility in the postwar era, several important strides were taken toward the production of steel in the South, and yet the enterprise ultimately failed.

The rolling mill was purchased just after the war by northern capitalists and incorporated as the Southwestern Iron Company in 1866. Puddling furnaces installed in the plant in 1869 permitted the company to roll fresh rail as well as reroll old rail. The plant was purchased in 1870 by J. T. Wilder and Hiram Chamberlain's successful Roane Iron Company, providing the mill with a reliable source of pig iron to convert into rail.

During the 1870s, however, railroads throughout the nation began laying steel rail in place of the less durable iron. Because of a declining market for its iron rail, the rolling mill halted operation in 1877. The survival of the mill depended on adoption of steel-making technology. New equipment was installed at the plant, and steel rail was rolled at the plant in 1878, the first such activity in the South. Maintaining a low phosphorus content in the steel was found to be difficult, and competition with northern rail again closed the plant. The company then installed new puddling equipment and a five-ton Bessemer converter. In 1887 the rolling mill produced the first Bessemer steel in the South.

The facility was purchased by the Southern Iron Company of Nashville in 1889 and was remodeled yet another time. Using ten-ton basic open-hearth steel furnaces, the rolling mill produced, in 1890, the first basic steel made in the South. The string of achievements at the rolling mill came to a permanent close in the same year when it was shut down for the last time; the plant's output was not competitive with northern steel. Despite continual technological innovation and adaptation to market demands, the rolling mill was an economic failure.

Table 6.1

Pig-Iron Production in the United States, 1876, by State

State or Territory	Blast Furnaces	Net Tons, Pig Iron	% of U.S. Total
Maine	1	3,002	0.14
New Hampshire	0	0	-
Vermont	2	550	0.03
Massachusetts	6	5,040	0.24
Rhode Island	0	0	-
Connecticut	10	10,160	0.49
New York	57	181,620	8.68
New Jersey	18	25,349	1.21
Pennsylvania	279	1,009,613	48.23
Delaware	0	0	-
Maryland	24	19,876	0.95
Virginia	33	13,046	0.62
North Carolina	8	400	0.02
Georgia	11	10,518	0.50
Alabama	13	24,732	1.18
Texas	1	426	0.02
West Virginia	12	41,165	1.97
Kentucky	23	34,686	1.66

Table 6.1, continued

State or Territory	Blast Furnaces	Net Tons, Pig Iron	% of U.S. Total
Tennessee	**24**	**24,585**	**1.17**
Ohio	100	403,277	19.27
Indiana	9	14,547	0.70
Illinois	12	54,168	2.59
Michigan	34	95,177	4.55
Wisconsin	14	51,261	2.45
Minnesota	1	0	-
Missouri	19	68,223	3.26
Oregon	1	1,750	0.08
California	0	0	-
Utah Territory	2	65	*
Wyoming Territory	0	0	-
Total	714	2,093,236	99.56

* Less than 0.01 percent (0.003).

Source: Swank, *Introduction to a History of Ironmaking*, 104.

The story of East Tennessee iron in the late nineteenth century was one of widespread business failures and lost investments.[29] Few companies enjoyed long periods of profit making, and most fell prey to one of the general economic depressions of the late nineteenth and early twentieth centuries. In the post–Civil War era in Tennessee the development of the coal and iron resources of the state was seen as the principal route to economic rejuvenation, and the involvement of northern and foreign entrepreneurs was enthusiastically cultivated.[30] Despite technological innovations and improved production statistics, the iron industry of the Chattanooga region was dwarfed by that of the Birmingham, Alabama, region. The thin veins of hematite iron ore in the Dyestone Iron Region of East Tennessee could not be exploited at an economically advantageous scale, and the high phosphorus content made its range of uses limited.[31]

One of the chief limitations of southern iron production in the nineteenth century had been the lack of local markets for its products. The agricultural base of the antebellum South could not absorb the potential output of regional furnaces; most furnaces operated at low, inefficient levels of production. Access to the growing industrial northern markets was difficult; the North had its own vigorous iron production. Further, southern iron tended to have a higher phosphorus content, making foundry-grade iron too brittle for many applications. In the late 1800s, a boon was finally experienced in the form of an emerging home market for cast-iron pipe.[32]

Developments in urban sanitation created a new market for cast-iron sewer pipe and pressurized water pipe. The creation of municipal sanitary sewerage and city water supplies in cities throughout the South beginning during the last quarter of the nineteenth century opened a new regional market for brittle, high-phosphorus southern iron.[33] The first plant built specifically for the founding of cast-iron pipe was erected at Chattanooga in 1877 and enlarged in 1884.[34] The production of water, sewer, and gas pipe steadily expanded during the late nineteenth and early twentieth centuries, particularly in Alabama. In 1898, pipe shops in Chattanooga, South Pittsburg, Anniston, and Bessemer were consolidated into the American Pipe and Foundry Company, which in turn consolidated with several northern plants in 1899 to form the United States Cast Iron Pipe and Foundry Company.[35] This firm's Chattanooga plant was still in operation in 1991.

Antebellum Industrialization in the South

Bluff Furnace and the company foundry at the Chattanooga rail yards were built by a corporation of southern industrialists and entrepreneurs in the midst of an economic and social system based on slave labor. As slave owners, Cravens and Whiteside were familiar with the advantages and limitations that this system entailed. Traditionally oriented charcoal-iron plantations in the South had always used slave labor, but in an orchestrated seasonal cycle linked with subsidiary agricultural operations. Slave-labor gangs were often leased to blast furnaces from local plantations on a seasonal basis. Charcoaling operations, for example, were conducted during the winter, when agricultural tasks were limited. Blasts were usually run after spring planting and before fall harvests.[36] When Henderson and company leased the plant in 1859, they found themselves strangers in a strange land, unaccustomed to a captive work force, and trying to impose sophisticated industrial technology onto a region overwhelmingly committed to agricultural pursuits. The nonseasonal free-labor work regimen with which the northerners were accustomed meshed poorly with those employed at southern iron plantations.

Agricultural interests dominated the economy of most southern states in the antebellum period. The largest industries in antebellum Tennessee were agriculture and two subsidiary industries: flour and grist milling, and lumber production.[37] This agrarian economic posture, and the overall record of Southern industry in the Civil War, contributed to the perception of the South as technologically backward. The lack of diversification into manufacturing and industrial pursuits was a fatal economic flaw. Modern scholars have reexamined the process of industrialization in the South and at least partially demolished the stereotype of the aristocratic planter society.[38] The principal conclusion of much of this research is that the South was not industrially *un*developed but *under*developed.

A review of current hypotheses about industrialization in the antebellum South reveals that one essential element retarding industrial growth was a reluctance of southern capitalists to invest in industry. Similarly, northern capitalists shied away from economic commitments in the industries of the South. Industrial investment was minimal.

The southerners most capable of investment in industrial pursuits were members of the planter class, usually defined as composed of individuals owning twenty or more slaves. In this group was concentrated capital, education, entrepreneurial ability, and political clout. Yet, as a group, only 6 percent of the planter class participated to any degree in manufacturing pursuits.[39] In a recent study of Tennessee entrepreneurs, two distinct groups of antebellum industrialists in the state were characterized: conservative, family-oriented manufacturers who worked mainly to consolidate their interests and secure family wealth, and more entrepreneurial, speculative investors who sought to maximize profits by investing across a wide spectrum of potentially profitable enterprises.[40] James A. Whiteside was a representative of the second, more entrepreneurial group; he invested in diverse industrial and commercial enterprises in Chattanooga and the surrounding region.

The unique aspect of southern iron production, and most other industries in the region, was the reliance on slave labor; the larger economic and social commitment of southern society to slavery welded slaves and the production of iron together.[41] One of the principal hypotheses to explain the backwardness of industry in the antebellum South was that slave labor inhibited the development of manufacturing in several ways. As perceived at the time, the purchase of slaves absorbed a lot of capital; slaves were unsuitable as industrial workers; white laborers did not work well next to slaves; immigration of skilled foreign and northern technicians was discouraged by the presence of slavery; and there was a fear of violent slave insurrection.[42]

In the South, these factors were interwoven in a peculiar and ironic pattern. There was the perception among planter/capitalists that slave labor was not adaptable to technological innovation, that in sophisticated technical tasks slave labor could not be effectively employed. This unadaptability was often imputed to the ignorance and ineducability of the slave, but more modern surveys of corporate histories agree that slave laborers were, in fact, capable of being trained to operate increasingly complex equipment and processes.

Planters who declined to introduce technological innovations in their agricultural production often did so for reasons not related to the skill potential of their slave work force; work was also an instrument of social control.[43] The factory was perceived as a much more difficult environment than the field in which to maintain discipline among the slave work force. Moreover, slaves idled by technological innovation were poor investments to the slave-owner/capitalist. Innovations that might have resulted in cheaper, and thus more competitive, pig-iron prices, were shunned if they disrupted the discipline of the work gang.[44] The irony here is that the use of slave labor was the economic edge that allowed southern ironmasters to produce pig iron as cheaply as they did, and substantial profits were, in fact, made at many facilities.[45] In short, southern planters and industrial entrepreneurs were not so much committed to low levels of technology as they were wedded to production systems that were labor intensive and gang oriented.

Modern analyses conclude that investment in southern industrial enterprise was profitable, in some cases more so than agricultural pursuits, but that for real or imagined reasons northern and southern investors failed to exploit opportunities in the South despite their profitability.[46] Nor did most southern firms expand to the degree that would allow them to benefit from internal economies of scale.[47] Some analyses of antebellum underindustrialization in the South suggest that the agrarian economy of the region was a rational economic response to the exploitation of a comparative advantage: agriculture was the most cost-effective form of investment in the region. The meteorological climate of the South may have been suited to agricultural production, but, as we have suggested, slave labor and agriculture meshed well; their interdependence maintained the southern *social* climate.

At Bluff Furnace in November 1860, with the controversy of Lincoln's election stirring the populace, relations between the northern plant managers and the southern furnace owners must have been tense. Likewise, the free-white labor force and the black slave-labor gangs stood at odds. In this atmosphere, inattention to routine blast-furnace duties could easily have caused the final blast failure. Technological innovations at the furnace were not independent of the human element in the equation.

Bluff Furnace: Past, Present, and Future

Even in its present configuration as an excavated archaeological site, Bluff Furnace is still embedded in a social, political, and economic matrix. The site is strategically located adjacent to a major riverfront redevelopment project in downtown Chattanooga. Developers, politicians, and planners with an eye to the future are beginning to take the site's past into account. By virtue of its location and its connection to Chattanooga's early history, the furnace is seen as having a modern "utility"; only now the function of the site has shifted from the production of durable goods to the creation and confirmation of ideas about modern life, how we view ourselves with reference to the past. In other words, Bluff Furnace has an educational role.

Chattanooga is blessed with a rich diversity of prehistoric archaeological sites spanning thousands of years, as well as numerous historic sites and parks relating to critical battles of the Civil War. By contrast, Bluff Furnace is a relatively modern industrial site, and no dramatic conflict between armies of blue and grey occurred there. Even without the dignified historical patina that comes with great age, famous personages, or momentous events, however, the furnace is still the focus of keen local interest.

No archaeological fieldwork has been undertaken there for years, and no interpretive exhibits have been developed as yet, but Bluff Furnace has continued to draw a steady stream of interested visitors. Despite the absence of the usual visitor amenities, the site has a powerful attraction for schoolchildren, tourists, and local residents precisely because it is an authentic reminder of the city's earliest days. It forms a direct link to the past in two ways. First, it contains tangible material remains that are *genuine*: the casting-shed foundations, the cupola base, the drilling scars on the bluff face—all are the result of the effort of people who actually lived and worked in Chattanooga during the 1850s, not some modern recreation or copy concocted solely to attract tourist dollars. These authentic remains can be directly experienced through the senses of sight and touch, and the experiences they produce cannot be duplicated at any other place or in any other way. Second, the site represents a starting point, at least in this region, for the industrial development of an essential facet of modern life, the production of iron.

Beginning with the W&A Railroad, industry has always played a critical role in Chattanooga's development, and while that role has diminished in recent years, it still contributes importantly to the city's economic productivity. Bluff Furnace adds historical depth and significance to part of a modern city's life and, in so doing, validates the lives and livelihoods of its residents.

Historical parks, museums, and exhibits play a unique role in the interpretation of the past.[48] Visitors are particularly receptive in the special environments of a museum gallery, reconstructed house, or restored archaeological site. In recreating the past, the historical reconstruction transports the visitor to a different age, where the lessons of history are eagerly and more readily learned, as long as those lessons are not excessively unpleasant. In these amenable learning situations, the archaeologist or historian invariably presents to the willing recipient a series of validations, justifications, and confirmations of the present. The past must always be interpreted within a framework, and that framework is derived from the commonly accepted values and norms of the larger society. If certain things today are different from 150 years ago, it is because of the inexorable "march of progress"; if things are still the same, it is because traditional values are "timeless." In either case, visitors learn less about the separate reality of the past than they do about the current social, economic, and ideological realities of the present. For the young, this process involves instruction on themes and beliefs that are held to be important in today's society. For the old, this means that what is already known and accepted is given historical validation.

There is nothing inherently wrong in the use of the past as a mirror for the present, as long as that function is recognized and historical accuracy is not sacrificed at the altar of ideological expediency.[49] Besides, the educational role is inevitable in the very act of interpretation, beginning with what is chosen for presentation by the historian/archaeologist. The question really is how the educational and socialization process is to be carried out. We argue that the most effective environment in which this process may occur is one that is believable to the visitor. A site that accurately presents genuine, tangible historical remains will create this environment in a way that an artificial recreation can never match. That is the strength of the Bluff Furnace site in its current role. It possesses the quality of authenticity that is necessary to the successful socialization/education function that now drives it. This quality cannot be manufactured; it simply exists by virtue of several fortuitous accidents of history. As archaeologists, all we can do is reveal as much as we can of what is already there. That is what we have attempted to do in this book.

Epilogue

The Bluff Furnace site is quiet again, this time for a period just now exceeding eight years. The processes that we interrupted briefly in the early 1980s continue unabated; from neglect and malice, the site slowly decays. Vandals have spray-painted the gray stone walls of the casting shed, and street people occasionally burn the timber used to revet the brick wall south of the furnace base. Parties unknown have scavenged metal from the site to sell for scrap, and vegetation and winter ice eat away at the stonework left exposed in anticipation of restoration or reconstruction work. The site awaits its rebirth.

Pausing to sit on some portion of the furnace foundations, one can take in the view north toward the river and imagine the site in the summer of 1860. There is the acrid smell of burning coke hovering over the charging deck, and, in the rutted road west of the furnace, dust and the fetid smell of mule dung mingle. The screeching blast is deafening near the casting shed, drowning out the braying animals and shouting workmen. A plume of smoke jets from the top of the smokestack, and the damper plate on the hot-blast stove bounces up and down in an uneven cadence. Sweating workmen move wheelbarrows across the wooden charging deck, and down the riverbank cascades a smoking flurry of discarded slag. As the roustabouts fill a barge with pig-iron bars, deckhands on the company steamboat stare at Bluff Furnace in awe.

Awakening from this revery, one takes in the sights, smells, and sounds of the place as it is today. The riverbank is lined with trees, not slag. A bass boat screams upriver at indecent speed past a lumbering barge and tug, and in the near river channel, a rowing shell glides noiselessly through the shadow of the bluff. Perched on a rock in the shallow water near the shoreline, a fisherman threads another fish on his catch string and bates his hook again. Atop the bluff, the groundskeepers at the Hunter Museum start up lawn mowers, and the smell of fresh-cut grass permeates the air. The traffic sounds of Riverfront Parkway flood the bluff in waves.

Once in a while there is a calm, when only the air and the birds move. Despite changes wrought by people and nature, the bluff holds a strong sense of presence and place. Standing near the ruined limestone walls of the casting shed on the furnace terrace high above the river, we are surrounded by history and modern monuments of civil engineering. To the west stands a high, pale-gray limestone pier supporting the delicate steel trestle of the Walnut Street Bridge, built in 1891. To the north is the wide Tennessee River, the avenue of commerce and history, calmed by Nickajack Dam downstream. MacClelland Island, stripped of all vegetation during the Civil War, is now forested again and is a wildlife sanctuary supervised by the Audubon Society. To the east is the Hunter Art Museum, perched at the very edge of the high river bluffs commanding a view of the river valley.

Public planners with the Tennessee Riverpark development now include some vision of the Bluff Furnace Historic Park in their own "artist's interpretation" of a downtown waterfront park. Connected with Ross's Landing by a boardwalk, Bluff Furnace will transport visitors to a remote past where antebellum ironmasters took a bold step toward the future and where the tragedy of the Civil War left its mark on Chattanooga's earliest heavy industry.

We hope that the archaeological site and its significance will not slip back into oblivion again. To the history of Bluff Furnace at Chattanooga we have contributed our own blood, sweat, and tears, and invested the place with special meaning. As archaeologists, we are continually made aware that all things material will, in time, revert back to the earth. Even iron will eventually decay and return to forms reminiscent of the ore from which it originated; the iron has only been borrowed from nature for a span of years. This book has attempted to salvage from decaying physical remnants the imperishable story of an enterprise in iron on the banks of the Tennessee River.

Appendices

Appendix 1

Tennessee blast furnaces enumerated in Lesley's *Iron Manufacturer's Guide* (1859)

East Tennessee

Furnace	Furnace Name	County	Built	Remodeled	Abandoned	Blast[a]	Power[a]	Activity
256	Independence	Johnson	—	—	1845-47	—	—	Abandoned
257	Amanda	Sullivan	1837	—	1837	—	—	Abandoned
258	Franklin	Sullivan	1838	—	—	Cold	Water	Active
259	Holston or Welcher's	Sullivan	1838	—	—	Cold	Water	Active
260	Unknown	Sullivan	—	—	1837	—	—	Abandoned
261	Unknown	Sullivan	—	—	1837	—	—	Abandoned
262	Union	Carter	1847	1855	—	Cold	Water	Active
263	Evelina	Carter	1835	—	1847	—	—	Abandoned
264	Aerial	Carter	1818	—	1847	—	—	Abandoned
265	O'Brien's	Carter	1836	—	1838	—	—	Abandoned
266	White's	Carter	1810	—	1845-47	—	—	Abandoned
267	Little Troublesome	Carter	1839	—	1842	—	—	Abandoned
268	Rockbridge	Carter	1840	—	1845	—	—	Abandoned
269	Pleasant Valley	Washington	—	—	—	Cold	Water	Active

	Name	County	Date			Status
270	Clark's Creek	Washington	—	—	—	Dilapidated[b]
271	Bright Hope	Greene	—	Before 1837	—	Abandoned
272	Legion	Cocke	1807	1827	—	Abandoned
273	Love's	Sevier	1837	1852	—	Abandoned
274	Ball Play	Monroe	1851	—	Cold	Water	Inactive since 1854[c]
275	Tellico	Monroe	1840	—	Hot	Water	Inactive since 1856[c]
276	Cumberland Gap	Claiborne	—	—	Cold	Water	Active
277	Belleville	Claiborne	1828	—	Cold	Water	Idle 1853-57
278	Speedwell	Campbell	1815	1830	—	—	Abandoned
279	Sharp's	Granger	—	1845	—	—	Abandoned
280	Miller's	Union	1837	—	Hot	Water	Active
281	Eagle	Roane	1839	—	Hot	Water/Steam	Active
282	Eagle No. 2	Roane	1844	1845	—	—	Abandoned
283	Piney Grove	Roane	1823	1828	—	—	Abandoned
284	Bluff	Hamilton	1854	—	Hot	Steam	Active

[a] Blast and power source for some abandoned furnaces not specified, but assumed to be cold blast and water powered.
[b] Stack considered in usable condition.
[c] Plants still considered in operation though idle.

Middle Tennessee

Furnace	Furnace Name	County	Built	Remodeled	Abandoned	Blast[a]	Power[a]	Activity
564	Saline	Stewart	1853	—	—	Cold	Steam	Idle since 1854
565	Great Western	Stewart	1854	—	—	Cold	Steam	Idle since 1856
566	Iron Mountain	Stewart	1854	—	—	Cold	Steam	Idle since 1855
567	Peytona	Stewart	1847	1856	—	Cold	Steam	Active
568	Clark	Stewart	—	—	—	Cold	Steam	Idle since 1857
569	Lagrange	Stewart	—	1857[d]	—	Hot/Cold	Steam	Active in 1857
570	Eclipse	Stewart	—	—	—	Cold	Steam	Active in 1857
571	Crosscreek	Stewart	1853	—	—	Cold	Steam	Active
572	Rough-and-Ready	Stewart	1850	—	—	Cold	Steam	Idle since 1856
573	Bellwood	Stewart	—	—	—	Cold	Steam	Active in 1857
574	Bear Spring	Stewart	—	—	1854	Cold	Steam	Abandoned
575	Dover No. 2	Stewart	1854	—	—	Cold	Steam	Active
576	Ashland	Stewart	1851	—	—	Cold	Steam	Active
577	Union	Montgomery	1853	—	1854	Cold	Steam	Abandoned
578	Blooming Grove	Montgomery	1834	—	1847-49	Cold	Steam	Abandoned
579	Poplar	Montgomery	—	—	—	Cold	Steam	Active in 1855
580	Yellow Creek	Montgomery	1802	—	—	Cold	Steam	Active
581	Sailor's Rest	Montgomery	1854	—	—	Cold	Steam	Active
582	Montgomery	Montgomery	—	—	—	Cold	Steam	Active in 1857
583	Phoenix	Montgomery	—	—	1854	Cold	Steam	Abandoned
584	Antonio O. K.	Montgomery	—	1857	—	Cold	Steam	Active

ID	Name	County	Year	Year2	Blast	Power	Status
585	Louisa	Montgomery	—	—	Cold	Steam	Active in 1855
586	Washington	Montgomery	—	c. 1839	Cold	Steam	Abandoned
587	Mount Vernon	Montgomery	1838	1839	Cold	Steam	Abandoned
588	Lafayette	Montgomery	—	—	Cold	Steam	Abandoned
589	Tennessee	Montgomery	—	1851	Cold	Steam	Abandoned
590	Cumberland	Dickson	1790	—	Cold	Steam	Active
591	Carroll	Dickson	1853	—	Cold	Steam	Active in 1857
592	Bellevue (Mammoth)	Dickson	1825	1834	Cold	Steam	Abandoned
593	Worley	Dickson	1844	—	Cold	Steam	Active
594	Piney	Dickson	1832	—	Cold	Steam	Active
595	Laurel	Dickson	1815	1856	Cold	Steam	Abandoned
596	Jackson	Dickson	1833	1854	Cold	Water	Abandoned
597	Oakland	Hickman	1854	—	Cold	Steam	Active
598	Aetna	Hickman	1846	—	Cold	Steam	Active
599	Cedargrove	Perry	—	—	Hot	Steam	Active in 1857
600	Cedargrove No 2	Perry	—	—	Hot	Steam	Active in 1857
601	Brownsport	Decatur	1848	—	Hot	Steam	Active
602	Decatur	Decatur	1854	—	Hot	Steam	Active
603	Marion	Hardin	—	—	Cold	Steam	Active in 1854
604	Forty-Eight No. 1	Wayne	1846	—	Cold	Steam	Active
605	Forty-Eight No. 2	Wayne	1846	—	Cold	Steam	Active

[a] Blast and power source for some abandoned furnaces not specified but assumed to be cold blast and water powered.
[b] Stack considered in usable condition. [c] Plants still considered in operation though idle. [d] Converted from hot to cold blast in 1857.

Appendix 2

Individuals Associated with Mechanical, Mining, and Iron Manufacturing Trades, from the 1860 U.S. Census, Hamilton County

House	Family	Name	Age	Occupation	Birthplace	District
436	437	Alexander Hickman	55	Mechanic	Virginia	15 Ooltewah
689	689	Jackson Bricklew	26	Machinist	South Carolina	14 Chattanooga
710	711	R. W. Mellington	29	Machinist	Pennsylvania	"
776	779	Robert Cravens	56	Iron Master	Tennessee	"
781	785	John Ladger (sp?)	19	Machinist	Cuba	"
"	"	William R. Jones	25	Machinist	New York	"
789	793	M. R. Metz (sp?)	33	Mining	Germany	"
801	804	Giles S. Edwards	35	Manager of Furnace	England	14 Chattanooga
815	817	Thomas Hearris	29	Mining	England	12 Soddy
816	818	James Henderson	30	Proprietor of Firm	New Jersey	14 Chattanooga
828	830	Abe Abrahams	30	Machinist	South Carolina	"
904	904	George Mills	27	Iron Moulding	England	"
933	933	Patrick Nelligan	25	Moulder in Foundry	Ireland	"
"	"	Patrick Henson (sp?)	26	Machinist	"	"
936	936	Andrew Waren	37	Repairing Cars [1]	Ireland	"
949	948	John Lincoln	17	Machinist Apprentice	Pennsylvania	"
954	954	Thomas Webster	40	Master Machinist	England	"
955	955	R. D. Mann	35	Machinist	Pennsylvania	"

957	957	Edward Adcock	18	Moulder in Foundry	North Carolina	"
960	960	Henry Wormleighton	36	Moulder	England	"
961	961	Charles Chapman	30	Machinist	England	"
962	962	William Ramsey	18	Machinist Apprentice	North Carolina	"
989	989	Claborn Bell [2]	26	Machinist	Tennessee	"
1663	1665	John Jones	30	Mining	Wales	4 Chattanooga
"	"	Michail Peters	22	"	Ireland	"
"	"	James Harris	33	"	Wales	"
"	"	David Harris	24	"	"	"
"	"	Joseph Reece	33	"	"	"
"	"	John Owens	55	"	"	"
"	"	John Davis	54	"	"	"
1664	1666	Alfred Jenkins	50	"	England	4 Chattanooga
1665	1667	John Camelton (sp?)	36	Miner	Scotland	4 Chattanooga
"	"	Thomas Phipps	39	"	England	"
"	"	Reubin Poole	33	"	"	"
2115	2117	David Fritz	36	Mining	Nor Noted	11 Sale Creek
2183	2185	William Mitchell	28	Miner	England	12 Soddy
"	"	Peter Lloyd	18	"	"	"
"	"	Richard Edwards	19	"	"	"
"	"	L. D. Ellis	23	"	Tennessee	"

Source: Eighth Census of the United States, 1869; Work Sheets, Schedule 1, Free Inhabitants, Hamilton County, Tennessee.

[1] Occupation unclear; carpenter, machinist or foundryman? [2] Free Mulatto.

Appendix 3

Chemical Analysis of Miscellaneous Nineteenth-Century Blast Slag Samples

Specimen Number	Constituent (Percentage by Weight)							Desulfurization Index $\dfrac{CaO+MgO}{SiO_2+Al_2O_3}$	Refractory Index $\dfrac{Al_2O_3}{CaO+SiO_2}$
	SiO_2	Al_2O_3	CaO	MgO	MnO	S	FeO		
1	53.50	14.80	20.20	5.10	2.60	0.30	0.40	0.37	0.20
2	55.30	14.50	19.00	3.30	2.90	0.40	0.60	0.32	0.20
3	55.80	15.00	17.80	2.70	2.86	0.35	0.66	0.29	0.20
4	55.00	13.10	19.30	2.90	3.74	0.28	1.94	0.33	0.18
5	40.40	11.20	38.40	5.20	-	trace	3.80	0.84	0.14
6	39.95	17.41	29.64	6.47	0.91	1.60	0.24	0.63	0.25
7	40.20	16.45	30.00	7.29	0.84	1.20	0.57	0.66	0.23
8	41.64	13.20	35.91	4.21	0.74	1.00	0.11	0.73	0.17
9	42.94	16.29	31.10	4.16	0.51	0.90	0.34	0.60	0.22
10	41.11	13.45	29.82	4.75	0.66	0.60	6.44	0.63	0.19
11	37.84	13.20	20.68	2.93	0.80	0.39	20.83	0.46	0.23
12	51.84	15.21	21.80	4.82	1.16	-	3.73	0.40	0.21
13	40.60	16.80	32.20	-	-	-	10.40	0.56	0.23
14	43.20	12.00	35.20	4.00	-	-	4.20	0.71	0.15
15	58.40	9.10	25.40	2.49	2.10	0.08	1.86	0.41	0.11
16	58.00	7.42	16.10	1.58	4.64	0.05	11.90	0.27	0.10

Sample Descriptions:

Sample 1 from Eaton-Hopewell Furnace, charcoal-coal fuel mix; black glassy slag. Reported in White, "X-Ray Fluorescent Analysis," 187.

Sample 2 from Eaton-Hopewell Furnace, charcoal-coal fuel mix; green-turquoise slag. Reported in White, "X-Ray Fluorescent Analysis," 187.

Sample 3 from Eaton-Hopewell Furnace, charcoal-coal fuel mix; green glassy slag. Reported in White, "X-Ray Fluorescent Analysis," 187.

Sample 4 from Eaton-Hopewell Furnace, charcoal-coal fuel mix; black slag. Reported in White, "X-Ray Fluorescent Analysis," 187.

Sample 5 from Scotland, coke-fired furnace. Reported in Fairbairn, *Iron*, 62.

Samples 6-9 from Great Britain, coke-fired furnace. Reported in Fairbairn, *Iron*, 99.

Samples 10-11 from Great Britain, coke-fired furnace; black glassy slag. Reported in Fairbairn, *Iron*, 101.

Sample 12 from Europe, charcoal furnace. Reported in Overman, *The Manufacture of Iron*, 231.

Samples 13-14 from Europe, coke-fired furnace. Reported in Overman, *The Manufacture of Iron*, 231.

Sample 15 from Great Western Furnace, Stewart County, Tennessee, charcoal furnace; blue slag.

Sample 16 from Great Western Furnace, Stewart County, Tennessee, charcoal furnace; milky blue-green slag.

Notes

Chapter 1

1. In the United States, the *Journal of the Society for Industrial Archeology* (hereafter referred to as *IA*) appeared in 1975; in Britain the first edition of *Industrial Archaeology* was published in 1963.

2. Here we are following the definition proposed by Kenneth Hudson, *World Industrial Archaeology* (Cambridge: Cambridge Univ. Press, 1979), 2.

3. R. F. Tylecote, *A History of Metallurgy* (London: Metals Society, 1976).

4. D. L. Schroeder and K. C. Ruhl, "Metallurgical Characteristics of North American Prehistoric Copper Work," *American Antiquity* 33, no. 2 (1968): 162–69; Lewis R. Binford, "Archaeology as Anthropology," *American Antiquity* 28, no. 2 (1962): 217–25; John W. Rutledge and Robert B. Gordon, "The Work of Metallurgical Artificers at Machu Picchu, Peru," *American Antiquity* 52, no. 3 (1987): 578–94; Hellen P. Pollard, "The Political Economy of Prehispanic Tarascan Metallurgy," *American Antiquity* 52, no. 4 (1987): 741–52.

5. The military superiority of Cortés, de Soto, Pizarro, and others was considerably enhanced by, and would not have been possible without, the introduction of common European diseases for which Native Americans had little or no natural resistance.

6. Brian Fagan, *People of the Earth* (Boston: Little, Brown, 1986), 393, 419–22.

7. Peter Lane, *The Industrial Revolution: The Birth of the Modern Age* (New York: Barnes and Noble, 1978); see especially chapters 12, 13, and 14.

8. W. David Lewis, *Iron and Steel in America* (Greenville, Del.: Hagley Museum, 1976), 17.

9. E. N. Hartley, *Ironworks on the Saugus* (Norman: Univ. of Oklahoma Press, 1957), 3–20.

10. Paul F. Paskoff, *Industrial Evolution: Organization, Structure, and Growth of the Pennsylvania Iron Industry, 1750–1860* (Baltimore: Johns Hopkins Univ. Press, 1983), 2–3.

11. For instance, both the Baltimore Company of Maryland and the Principio Company of Virginia and Maryland were established in the second quarter of the eighteenth century.

12. Due to high capital costs, industrial machinery is not often left behind for the benefit of invading armies or future archaeologists. This was as true in 1863 as it is today.

13. Dr. Jeffrey Lawrence Brown, an energetic and innovative researcher, was from 1973 to 1980 a member of the faculty of the Department of Sociology and Anthropology at the University of Tennessee at Chattanooga (UTC). In his brief career he made a number of research contributions to industrial archaeology and was on the board of directors of the Society for Industrial Archeology. He died of cancer in 1980 at the age of forty. The following year, the UTC Institute of Archaeology was renamed in his honor.

14. David DeJarnette and Thomas H. DeJarnette, Jr., "Geology and Archaeology of the Old Tannehill Furnace Site" (Report, Tannehill Historical State Park, McCalla, Ala., 1956).

15. James R. Bennett, *Old Tannehill: A History of the Pioneer Ironworks in Roupes Valley, 1829–1865* (Birmingham, Ala.: Jefferson County Historical Commission, 1986).

16. Samuel D. Smith, Charles P. Stripling, and James M. Brannon, *A Cultural Resource Survey of Tennessee's Western Highland Rim Iron Industry, 1790s–1930s*, Research Series, no. 8 (Nashville: Tennessee Dept. of Conservation, Div. of Archaeology), 1988.

17. Les Crocker, "An Early Iron Foundry in Northern Mississippi," *Journal of Mississippi History* 35 (1973): 113–27.

18. John R. White, "X-Ray Fluorescent Analysis of an Early Ohio Blast Furnace Slag," *Ohio Journal of Science* 77 (1977): 186–88, and "Archaeological and Chemical Evidence for the Earliest American Use of Raw Coal as a Fuel in Ironmaking," *Journal of Archaeological Science* 5 (1978): 391–93.

19. Victor R. Rolando, "Ironmaking in Vermont: 1775–1890," Master's thesis, College of Saint Rose, Albany, N.Y., 1980).

20. Bruce E. Seely, "Blast Furnace Technology in the Mid-Nineteenth Century: A Case Study of the Adirondack Iron and Steel Company," *IA* 7, no. 1 (1981): 27–54.

Chapter 2

1. For an excellent history of iron manufacture in the eighteenth century, see James A. Mulholland, *A History of Metals in Colonial America* (University: Univ. of Alabama Press, 1981). The technology of iron smelting in the mid-nineteenth century was discussed in detail by a number of British and American ironmasters. Frederick Overman's *The Manufacture of Iron in All Its Various Branches* (Philadel-

phia: H. C. Baird, 1854), presents an American perspective on early blast-furnace technology and is written in a simple, comprehensible style. Akin to a technological encyclopedia, Andrew Ure's compendium, *A Dictionary of Arts, Manufactures and Mines; Containing a Clear Exposition of their Principles and Practice* (New York: D. Appleton and Co., 1868), vol. 1, contains much useful information on period smelting and related subjects; its perspective is European. British commentary on mid–nineteenth-century furnace practice was produced by William Fairbairn in his *Iron; Its History, Properties and Processes of Manufacture* (Edinburgh: Adam and Charles Black, 1865). Taking a more historical perspective was James Swank in *History of the Manufacture of Iron in All Ages*, 2d ed.(Philadelphia: American Iron and Steel Association, 1892). The bulk of this treatise was devoted to the development of the American iron industry, and Swank's commentary on Bluff Furnace is the linchpin in accounts of that operation. Two early twentieth-century American works are also useful in a synopsis of nineteenth-century blast-furnace operations. *The Metallurgy of Iron and Steel*, 2d ed. (New York: McGraw-Hill Book Co., 1913) by Bradley Stoughton and *The Principles, Operation and Products of the Blast Furnace* (New York: McGraw-Hill, 1918) by J. E. Johnson, Jr., put nineteenth-century practice in clearer perspective. These books clarify many of the chemical processes taking place in the furnace and document the result of developments in blast-furnace technology in the previous century.

2. Herman H. Chapman, *The Iron and Steel Industries of the South* (Birmingham: Univ. of Alabama Press, 1953), 27–28.

3. Johnson, in *The Principles, Operation and Products of the Blast Furnace*, 189, reported that 80 percent of the iron being smelted was from hematite ores.

4. The two principal varieties are goethite or hydrogen iron oxide $HFeO_2$, and lepidocrocite or monobasic ferric oxide, $FeO(OH)$.

5. Summarized from Ernest F. Burchard, *The Red Iron Ores of East Tennessee* (Nashville: Tennessee State Geological Survey, Bulletin 16, 1913; rpt. 1962), and Stuart W. Maher, *The Brown Iron Ores of East Tennessee* (Nashville: Tennessee Dept. of Conservation, Div. of Geology, Report of Investigations, no. 19, 1964).

6. Overman, *The Manufacture of Iron*, 39.

7. Ibid., 116.

8. Swank, *History of the Manufacture of Iron*, 352–65.

9. Johnson, *The Principles, Operation and Products of the Blast Furnace*, 170.

10. From statistics of the American Iron and Steel Association, quoted in Swank, *History of the Manufacture of Iron*, 376.

11. Seely, "Blast Furnace Technology," 27–54.

12. The shift from charcoal to coke iron production involved a number of factors, of which economics was but the central issue. The reputation of charcoal iron for certain uses, such as chilled car wheels for railroads, often outlived any real chemical superiority over coke-produced iron. As metallurgical science, testing procedures, and chemical analysis techniques improved in the nineteenth century, iron came to be ordered not on the basis of its reputation but on its metallurgical analysis. This done, the charcoal iron industry faded into oblivion, the cost of its product being higher than coke iron. See Robert B. Gordon, "Materials for Manufacturing: The Response of the Connecticut Iron Industry to Technological Change and Limited Resources," *Technology and Culture* 24 (1983): 602–34; Richard Shallenberg, "Evolution, Adaptation, and Survival: The Very Slow Death of the American Charcoal Iron Industry," *Annals of Science* 32, no. 4 (1975): 341–58; Richard Shallenberg and David A. Ault, "Raw Materials Supply and Technological Change in the American Charcoal Iron Industry," *Technology and Culture* 18 (1977): 436–66.

13. One indication of the capacity of a nineteenth-century furnace was to state its width "at the boshes."

14. The unaccented, anglicized spelling, *tuyere*, is used here.

15. Fairbairn, *Iron*, 59.

16. Overman, *The Manufacture of Iron*, 162–63.

17. Victor S. Clark, *History of Manufactures in the United States, Volume II 1860–1893* (New York: Peter Smith, 1949; rpt. of 1929 ed.), 76.

18. In "Blast Furnace Technology," Seely provides an excellent illustration of a water-powered bank of horizontal-cylinder blast pumps.

19. Clyde A. Sanders and Dudley C. Gould, *History Cast in Metal: The Founders of North America* (Des Plaines, Ill.: Cast Metals Institute of the American Foundrymen's Society, 1976), 30; Fairbairn, *Iron*, 59.

20. The blast delivered into the hearth was always fresh, oxygenated air, not exhaust gases from the furnace top; only the *heat* of the waste gases was recycled.

21. Overman, *The Manufacture of Iron*, 441–42.

22. Ibid., 435.

23. Swank, *History of the Manufacture of Iron*, 352–65.

24. The term *ironmaster* evokes images of a master craftsman or artisan rather than a technocrat or engineer.

25. Johnson, *The Principles, Operation and Products of the Blast Furnace*, 139–57.

26. Overman, *The Manufacture of Iron*, 153.

27. Johnson, *The Principles, Operation and Products of the Blast Furnace*, 229.

28. Overman, *The Manufacture of Iron*, 229–30.

29. Johnson, *The Principles, Operation and Products of the Blast Furnace*, 157.

30. Overman, *The Manufacture of Iron*, 401.

31. Ure, *A Dictionary of Arts, Manufactures and Mines*, 1069–70.

32. Ibid., 1073. We have converted Ure's measures in tons using 2,000 pounds as the equivalent; a long ton measure of 2,240 pounds may be more correct.

33. Overman, *The Manufacture of Iron*, 203.

34. Crushed blast slag made excellent road surfacing material as it was hard and durable, and water drained through it readily. In the late nineteenth century, blast slag was added as a strengthening aggregate to concrete. Certain types of slag were also used at puddling furnaces and rolling mills in the production of wrought iron, and the lime content of Bessemer slags made them excellent fertilizers after crushing. Stoughton, *The Metallurgy of Iron and Steel*, 108.

35. Lewis, *Iron and Steel in America*, 52–54.

Chapter 3

1. The iron-ore and coal deposits of Tennessee are described in detail in the following works: W. S. Bayley, *The Magnetic Iron Ores of East Tennessee and Western North Carolina* (Nashville: Tennessee Dept. of Education, Div. of Geology, Bulletin 29, 1923); Burchard, *The Red Iron Ores of East Tennessee*; Ernest F. Burchard, *The Brown Iron Ores of the Western Highland Rim, Tennessee* (Nashville: Tennessee Div. of Geology, Bulletin 39, 1934); J. B. Killebrew, *Iron and Coal in Tennessee: Report by the Commissioner of Agriculture, Statistics and Mines* (Nashville: Tennessee Bureau of Agriculture, 1881); Robert E. Hershey and Stuart W. Maher, *Limestone and Dolomite Resources of Tennessee* (Nashville: Tennessee Div. of Geology, Bulletin 65, 1963); John Rogers, *Geology and Mineral Deposits of Bumpass Cove, Unicoi and Washington Counties, Tennessee* (Nashville: Tennessee Dept. of Conservation, Div. of Geology, Bulletin 54, 1948); Wilbur A. Nelson, *The Southern Tennessee Coal Field* (Nashville: Tennessee Dept. of Education, Div. of Geology, Bulletin 33-A, 1925); Edward T. Luther, *The Coal Reserves of Tennessee* (Nashville: Tennessee Div. of Geology, Bulletin 63, 1959); James M. Safford, *Geology of Tennessee* (Nashville: S. C. Mercer, 1869); James M. Safford and Joseph B. Killebrew, *The Elements of the Geology of Tennessee* (Nashville: Ambrose and Bostelman, 1904).

2. See James Larry Smith, "Historical Geography of the Southern Charcoal Iron Industry, 1800–1860," Ph.D. thesis, Univ. of Tennessee, Knoxville, 1982.

3. Swank, *History of the Manufacture of Iron*, 299.

4. Raymond F. Hunt, "The Pactolus Ironworks," *Tennessee Historical Quarterly* 25 (1966): 176–96.

5. Robert E. Dalton, "Montgomery Bell and the Narrows of Harpeth," *Tennessee Historical Quarterly* 35 (1976): 3–28. A 290-foot tunnel, cut by slave labor through solid rock, crossed a narrow neck of land where the Harpeth River looped back upon itself. The difference in water elevation across the Narrows of Harpeth provided fall for Bell's water-powered forges.

6. Discussed in E. Raymond Evans and Vicky Karhu, "Cultural Overview and Synthesis Study of the Chattanooga Riverfront, Chattanooga, Tennessee," report prepared for Moccasin Bend Task Force, City of Chattanooga and Hamilton County, 1985, 104–5.

7. Penelope Johnson Allen, "Leaves from the Family Tree: Cravens," *Chattanooga Times*, Mar. 18, 1934.

8. Eugene Monroe Pickel, *A History of Roane County, Tennessee, to 1860* (Kingston, Tenn.: Roane County Heritage Commission, 1981), 36.

9. Mary Snyder and Phil Noblitt, "Robert Cravens: A Brief History of the Man and His Family" (Chickamauga and Chattanooga National Military Park, 1975), 7.

10. J. Peter Lesley, *The Iron Manufacturer's Guide to the Furnaces, Forges and Rolling Mills of the United States* (New York: J. Wiley, 1859), 82–83.

11. Snyder and Noblitt, "Robert Cravens," 7.

12. Lesley, *The Iron Manufacturer's Guide*, 82–83.

13. Roane County Deed Book J: 337.

14. Lesley, *The Iron Manufacturer's Guide*, 204.

15. This was a problem first overcome by Abraham Darby in Coalbrookdale, England, as early as 1709. Two years before Darby began the first successful iron smelting with coked coal, he perfected molding techniques that permitted thin-walled castings to be made. Coke iron, being more fluid than charcoal iron, meshed well with the new molding technology. Darby can be credited with creating first a product for coke iron, then the coke iron itself. Charles K. Hyde, *Technological Change and the British Iron Industry, 1700–1870* (Princeton, N.J.: Princeton Univ. Press, 1977), 40.

16. Chattanooga's early railroad history is summarized by James W. Livingood in "Chattanooga: A Rail Junction of the Old South," *Tennessee Historical Quarterly* 6 (1947): 230–50.

17. John P. Long, one of Chattanooga's earliest inhabitants, recalled an early attempt to exploit the iron-making potential of the Chattanooga region: "The first effort looking to the iron manufacture was made about the year 1850. A Mr. Hollister, a practical ironmaster, visited this place about that time, made an examination of the ores and the coal around here, and was delighted with the prospect. He succeeded in raising a company with the necessary capital, went North and perfected his plans and specifications, and on his way out took sick and died at Charleston, which broke up the enterprise." "Early Days of Chattanooga," in *Transactions of the Iron, Coal and Manufacturer's Association of Chattanooga, Tennessee, for the Year 1880* (Chattanooga: Iron, Coal and Manufacturer's Association of Chattanooga, 1880), 18.

18. *Acts of the State of Tennessee passed at the First Session of the Twenty-Seventh General Assembly, for the years 1847–8* (Jackson, Tenn.: Gates and Parker, 1848), 47–48.

19. T. J. Campbell, *The Upper Tennessee: Comprehending Desultory Records of River Operations in the Tennessee Valley* (Chattanooga: pub. by author, 1932), 32–33.

20. Biographical data from Lucile Deaderick, ed., *Heart of the Valley: A History of Knoxville, Tennessee* (Knoxville: East Tennessee Historical Society, 1976). Swan donated land for a public market in Knoxville and in 1854 was awarded the franchise for the city's gas lighting. He was a steamboat investor and later a member of the Confederate Congress.

21. Thomas Clark Lyon was born in Roane County in 1810 and graduated East Tennessee University in 1829. Joshua W. Caldwell, *Sketches of the Bench and Bar of Tennessee* (Knoxville: Ogden Brothers and Co., 1898), 302–4.

22. James W. Livingood, *A History of Hamilton County, Tennessee* (Memphis: Memphis State Univ. Press, 1981), 111–12.

23. Roane County Deed Book L-1: 93.

24. Hamilton County Deed Book 10: 76.

25. F. N. Boney, "Part Three: 1820–1865," in *A History of Georgia*, ed. Kenneth Coleman (Athens: Univ. of Georgia Press, 1977), 166.

26. Now identified as Burns Island, situated just downstream from the mouth of the Sequatchie River.

27. Hamilton County Deed Book 10: 392; the deed of a town lot from the company to Thomas Patterson contains the entry, "Rec'd payment in full of the within described note and interest 18th day of March 1854 East Ten. Iron Mfg. Co. by S. C. Rogers."

28. Johnson traded land in Roane County for stock. In 1857, the administrators of Luke Lee (sometimes Lea), deceased, relinquished his twenty shares of stock in the company in exchange for Lot 5 on James Street, Chattanooga; Hamilton County Deed Book 12: 577.

29. Roane County Deed Book S-1: 107.

30. Ernest M. Lander, Jr., "Charleston: Manufacturing Center of the Old South," *Journal of Southern History* 26 (1960): 350. Ker Boyce died in 1854.

31. Ker Boyce was also a partner in another local company, although it existed only on paper. Section 5 of the Tennessee General Assembly act incorporating the East Tennessee Iron Manufacturing Company chartered a separate corporate entity, the Chattanooga Iron Manufacturing Company, consisting of subscribers George W. Crawford, Ker Boyce, Farish Carter, John P. King, J. Edgar Thomson, James A. Whiteside, Benjamin Chandler, Rease B. Brabson, Thomas G. McFarland, Robert M. Hooker, and Joseph P. McCollough. Despite the impressive array of capital, entrepreneurial skill, and political clout of the incorporators, there is no evidence that the Chattanooga Iron Manufacturing Company ever actually operated.

32. Cited in Ronald L. Lewis, *Coal, Iron, and Slaves: Industrial Slavery in Maryland and Virginia, 1715–1865*, Contributions in Labor History 6 (Westport, Conn.: Greenwood Press, 1979), 248. Facilities producing wrought iron (bloomaries, forges, and rolling mills) are not distinguished.

33. J. D. B. DeBow, *A Statistical View of the United States, . . . Being a Compendium of the Seventh Census* (Washington, D.C.: A. O. P. Nicholson, Public Printer, 1854), 181.

34. 1860 production statistics cited in Robert H. McKenzie, "Reconstruction of the Alabama Iron Industry, 1865–1880," *Alabama Review* 25 (1972): 179.

35. Cited in Burchard, *Brown Iron Ores of the Western Highland Rim*, 8.

36. Safford, *Geology of Tennessee*, 464–66.

37. Edward T. Luther, *Our Restless Earth: The Geological Regions of Tennessee* (Knoxville: Univ. of Tennessee Press, 1977), 35. The northwestern United States at that time included western Pennsylvania and Ohio.

38. This tract was purchased from Benjamin R. Montgomery; Hamilton County Deed Book 8: 209–10.

39. *Chattanooga Gazette*, Apr. 1853. Cited in Gilbert E. Govan and James W. Livingood, *The Chattanooga Country, 1540–1976: From Tomahawks to TVA* (Knoxville: Univ. of Tennessee Press, 3d ed., 1977), 165.

40. Nelson, *The Southern Tennessee Coal Field*, 4, also notes: "The Roddy banks in Rhea County, were another of the old producers. In 1853, about 900 tons [of coal]

were mined here and shipped to the Eagle Furnace at Chattanooga for metallurgical uses." Some confusion between Eagle Furnace in Roane County and Bluff Furnace in Chattanooga is evident. It is also unclear to what use the coal was put; in 1853, only the foundry of the East Tennessee Iron Manufacturing Company was in operation at Chattanooga.

41. Hamilton County Deed Book 8: 408.

42. *Acts of the State of Tennessee passed at the First Session of the Twenty-Ninth General Assembly, for the years 1851–2* (Nashville: M'Kennie and Brown, 1852), 270.

43. Chattanooga Island is now known as Maclelland Island.

44. Deed records for this period may be incomplete. Hamilton County Deed Book 9 (I) is not extant. The gap between Book 8 or H and 10 or J covers the period March 1852 to June 1853. Transactions discussed above and falling in this period were recorded in later books.

45. Lesley, *The Iron Manufacturer's Guide*, 83.

46. Hamilton County Deed Book 10: 62. Whiteside was acting on behalf of himself and Tomlinson Fort, John S. Thomas, Samuel Williams, J. P. Boyce, and Farish Carter, who had given him leave to conduct the sale of the property. The unspecified consideration might well have been stock.

47. Hamilton County Deed Book 10: 202–3.

48. Lesley, *The Iron Manufacturer's Guide*, 83.

49. Ibid.

50. East Tennessee blast furnaces are covered in Lesley, *The Iron Manufacturer's Guide*, 79–83; Middle Tennessee furnaces, 130–37.

51. Lesley's statistics on Tennessee iron facilities were current to 1857–58.

52. "Strother, David Hunter," in *Dictionary of American Biography*, ed. Dumas Malone (New York: Charles Scribner's Sons, 1936).

53. David Hunter Strother, engraving of Bluff Furnace, *Harper's New Monthly Magazine* 17 (1858): 298.

54. Southern bituminous coals were tarry and sulfurous and not well suited for this chemical reason when burned raw in blast furnaces. The process of coking improved the desirability of the coal as fuel.

55. *Acts of the State of Tennessee passed at the First Session of the Thirtieth General Assembly, for the years 1853–4* (Nashville: M'Kennie and Brown, 1854), 593–611.

56. David Rankin of Marion County, Tennessee, was born in 1799, elected to the state legislature in 1845 and reelected in 1847. Rankin was instrumental in the

construction of the Higginbotham Toll Pike in Marion County in 1844. He died in 1862 while serving as clerk and master of the Marion County Chancery Court at Jasper. Erasmus Alley was also a Marion County resident. Biographical data from J. Leonard Raulston and James W. Livingood, *Sequatchie: A Story of the Southern Cumberlands* (Knoxville: Univ. of Tennessee Press, 1974).

57. Marion County Deed Book G: 290–94.

58. Hamilton County Deed Book 12: 527–28.

59. Marion County Deed Book H: 390–91. The company mortgaged some of its mine equipment and facilities to A. C. Vaness to secure two notes. M. Pratt signed the mortgage as an agent of the Etna Mining and Manufacturing Company.

60. J. B. Killebrew, *Introduction to the Resources of Tennessee: First and Second Reports of the Bureau of Agriculture, State of Tennessee* (Nashville: Tavel, Eastman and Howell, 1874), 193.

61. Marion County Deed Book H: 790–93.

62. Ibid., 753–54.

63. Ibid., 561–63. The mortgage was given to Robert T. Andrews, trustee, in order to secure thirty notes of five hundred dollars each.

64. J. J. Ormsbee, "The Rise and Progress of Coal Mining in Tennessee," in A. D. Hargis, *Seventh Annual Report of the Bureau of Labor, Statistics and Mines of the State of Tennessee, for the year ending December 31, 1897* (Nashville: Breeder and Horseman, 1898), 21.

65. Henry M. Wiltse, *History of Chattanooga* (Chattanooga: pub. by author, 1916–19), vol. 1: 16–17, 71.

66. Hamilton County Deed Book 11: 376.

67. *The Chattanooga Gazette*, May 10, 1856.

68. Hamilton County Deed Book 13: 266.

69. Hamilton County Deed Book 21: 669. The bond for title was badly damaged during a fire, and it was not until 1871 that the final warranty deed was filed for record.

70. See note 44, above.

71. For example, Charles. D. McGuffey, ed., *Standard History of Chattanooga, Tennessee* (Knoxville: Crew and Dorey, 1911), 29.

Notes to pages 70–76 / 203

72. Major George B. Davis, Leslie J. Perry, and Joseph W. Kirkley, compilers, *The War of the Rebellion, A Compilation of the Official Records of the Union and Confederate Armies* (Washington, D.C.: Government Printing Office, 1890), ser. 1, vol. 31: 536.

73. Ethel Armes, *The Story of Coal and Iron in Alabama* (Birmingham, Ala.: Birmingham Chamber of Commerce, 1910; rpt. 1973, New York: Arno Press), 175.

74. Giles Edwards to John Fritz, Feb. 27, 1858, John Fritz Correspondence, Canal Museum, Hugh Moore Historical Park and Museums, Easton, Pa.

75. Original minutes of the city council have been lost. The city ordinance was recorded from secondary sources and recorded in Wiltse, *History of Chattanooga* 2: 55.

76. The photographic stereoscope is in the Dr. James Livingood Collection at the Chattanooga Regional History Museum and is reproduced here through Dr. Livingood's kindness. The stereograph is marked: "Tucker and Perkins, Southern Stereoscopic Views, Published and for sale at their Photographic Galleries, 192 Broad Street (South Side), Augusta, Georgia [and] Corner of Broad and Randolf Streets, Columbus, Georgia."

77. The Potomac Furnace in Loudon County, Virginia, commenced production of coke pig iron with a steam-driven hot-blast in 1848. Lesley, *The Iron Manufacturer's Guide*, 63.

78. John Fritz, *The Autobiography of John Fritz* (New York: J. Wiley and Sons, 1912), 135. This coal and ore property might have been the Etna Mines, under lease to the Raccoon Mountain Coal Company of New York.

79. Fritz, *Autobiography*, 141.

80. Abram S. Hewitt to John Fritz, Oct. 20, 1860, Fritz Correspondence.

81. U.S. Census Bureau, *Eighth Census of the United States, 1860*, Work Sheets, Population Schedule 1, Free Inhabitants, Hamilton County, Tenn.

82. James M. Swank, *Introduction to a History of Ironmaking and Coal Mining in Pennsylvania* (Philadelphia: pub. by author, 1878), 71.

83. "Jones, William Richard," in *Dictionary of American Biography*, ed. Malone (1933).

84. McGuffey, ed., *Standard History of Chattanooga*, 174–75.

85. Govan and Livingood, *The Chattanooga Country*, 169, point out that Harriet Lloyd is listed in enumerator's schedules of the 1860 census as being resident in Chattanooga with her parents. Jones and Lloyd were married on April 14, 1861, and apparently departed Chattanooga shortly afterward. Tennessee had already seceded from the Union by that date.

86. Armes, *The Story of Coal and Iron in Alabama*, 176.

87. J. Eugene Lewis, "Cravens House, Landmark of Lookout Mountain," *Tennessee Historical Quarterly* 20 (1961): 207.

88. This invisibility of blacks in the pages of history is a prime rationale for the emergence of archaeological studies of slavery and nineteenth-century black history.

89. Robert S. Starobin, *Industrial Slavery in the Old South* (New York: Oxford Univ. Press, 1970), 15.

90. Concentrations of slave laborers in the urban setting of the Tredegar works was a source of anxiety to townspeople more accustomed to the small numbers of black domestics employed in their households.

91. Charles B. Dew, "Black Ironworkers and the Slave Insurrection Panic of 1856," *Journal of Southern History* 41 (1975): 321–38; Harvey Wish, "The Slave Insurrection Panic of 1856," *Journal of Southern History* 5 (1939): 206–22.

92. Swank, *History of the Manufacture of Iron*, 290.

93. Ormsbee, "The Rise and Progress of Coal Mining in Tennessee," 23.

94. Recounted in the *Chattanooga Sunday Times*, July 26, 1914, 29.
Marcus B. Long was the son of John Pomfret Long, one of the first settlers at Chattanooga and a keen observer of the progress of the town. In another article (*Chattanooga Sunday Times*, Nov. 8, 1885, 11), J. E. McGowan states that a Connecticut man named Churchill ran the operation and abandoned the furnace when it chilled just prior to the war. McGowan also insisted that the iron ore was mined close to the city, on the north side of the river. The source of McGowan's information is unknown.

95. Armes, *The Story of Coal and Iron in Alabama*, 176.

96. Swank, *History of the Manufacture of Iron*, 290. The machinery was evidently shipped to the Oxford Furnace in Calhoun County. Bennett, *Old Tannehill*, 19, 83; Joseph H. Woodward II, *Alabama Blast Furnaces* (Woodward, Ala.: Woodward Iron Co., 1940), 105.

97. *Chattanooga News*, Aug. 27, 1891.

98. Gleason may conceivably be the Mr. Greacon referred to in the Edwards-Fritz correspondence noted above.

99. Hamilton County Deed Book 15: 151–52.

100. Kay Baker Gaston, "The Remarkable Harriet Whiteside," *Tennessee Historical Quarterly* 40 (1981): 337.

101. Hamilton County Deed Book 15: 529–31.

102. Govan and Livingood, *The Chattanooga Country*, 199–200.

103. Swank, *History of the Manufacture of Iron*, 290.

104. The photograph is now in the collection of the Jeffrey L. Brown Institute of Archaeology, University of Tennessee at Chattanooga.

105. "Coal Mining in Pioneer Days As Recalled by Col. 'Bill' Nixon," *Chattanooga Times*, Mar. 18, 1923.

106. Davis, Perry, and Kirkley, compilers, *The War of the Rebellion*, ser. 1, vol. 31: 536.

107. See James F. Doster, "The Chattanooga Rolling Mill: An Industrial By-Product of the Civil War," *East Tennessee Historical Society's Publications* 36 (1964): 45–55.

108. Marion County Deed Book H: 761–62.

109. Ibid., 755–59.

110. Brig. Gen. Fred C. Ainsworth and Joseph W. Kirkley, *The War of the Rebellion: A Compilation of Official Records of the War of the Rebellion* (Washington, D.C.: Government Printing Office, 1900), ser. 4, vol. 3: 291.

111. Discussed in Joseph H. Woodward II, "Alabama Iron Manufacturing, 1860–1865," *Alabama Review* 7 (1954): 201.

112. Richard D. Goff, *Confederate Supply* (Durham, N.C.: Duke Univ. Press, 1969), 58.

113. Woodward, "Alabama Iron Manufacturing," 201, 203–4.

114. J. W. Joseph and Mary Beth Reed, "Ore, Water, Stone and Wood: Historical and Architectural Investigations of Donaldson's Iron Furnace, Cherokee County, Georgia," report (Atlanta: Garrow and Associates, 1988), 54.

115. Woodward, "Alabama Iron Manufacturing," 206.

116. *Chattanooga Daily Rebel*, Oct. 7, 1862.

117. Woodward, "Alabama Iron Manufacturing," 204.

118. Henderson's patents were: 50,474, granted October 17, 1865, for the addition of manganese to iron being converted by the pneumatic Bessemer process to steel; 93,713, granted August 17, 1869, for an apparatus to be used in the decarburizing process of iron during steel-making;101,263, granted March 29, 1870, for the use of fluorspar and titaniferous iron ores in the manufacture of steel and malleable iron; 106,365, granted August 16, 1870, for treating crude iron with fluorides and oxides.

119. William Battle Phillips, *Iron Making in Alabama* (University: Geological Survey of Alabama, 3d ed., 1912), 9.

120. Stewart H. Holbrook, *Iron Brew: A Century of American Ore and Steel* (New York: Macmillan Co., 1939), 207–15, and Malone, ed., *Dictionary of American Biography* (1933), vol. 10: 208–10.

121. McKenzie, "Reconstruction of the Alabama Iron Industry," 185.

122. Armes, *The Story of Coal and Iron in Alabama*, 204, 212, 229–31.

123. American Iron and Steel Association, *Directory to the Iron and Steel Works of the United States* (Philadelphia, 1880), 51.

124. Lewis, "Cravens House," 204.

125. Jeffrey L. Brown, "The Camp Site Beneath the Cravens House Porch," *Tennessee Anthropologist* 3 (1978): 6, 8.

126. R. N. Price, *Holston Methodism, From Its Origin to the Present Time* (Nashville: Publishing House of the Methodist Episcopal Church, South, 1913), vol. 5: 337–38.

127. John Wilson, *Chattanooga's Story* (Chattanooga: Chattanooga News–Free Press, 1980), 166.

128. Lewis, "Cravens House," 207; Snyder and Noblitt, "Robert Cravens," 3.

129. In the flood commentary of the *Daily American Union* (Knoxville), the lime kiln at the bluff was mentioned: "At the river bank nothing could be seen but water outspread as far as the eye could reach. The current was still very swift and considerable drift wood was running down. The water was up to the base of the lime kiln on the cliff above town. This shows that the water has risen fifty feet, the base of the kiln being about that distance above the average stage of water." Quoted in TVA, *Floods on Tennessee River, Chattanooga and Dry Creeks, and Stringers Branch, Vicinity of Chattanooga, Tennessee* (Knoxville: TVA, Div. of Water Control Planning, 1959), 33.

130. Hamilton County Deed Book 48: 409.

131. Son of Dr. Tomlinson Fort, Col. Tomlinson Fort evidently inherited an interest in the furnace from his father.

Chapter 4

1. M. Elizabeth Will worked as a consultant for the institute, specializing in soil analysis; R. Bruce Council was hired in 1979 as the institute's research instructor, supervising excavations of the Union Railyards Site in downtown Chattanooga.

2. Ken Kesey, *Sometimes A Great Notion* (New York: Penguin Books, 1977).

3. The following winter, Honerkamp received great satisfaction from burning the offending tree, a little at a time, in his living-room fireplace.

4. R. Bruce Council, M. Elizabeth Will, and Nicholas Honerkamp, "Bluff Furnace: Archaeology of a Nineteenth Century Blast Furnace," manuscript, Jeffrey L. Brown Institute of Archaeology, 1982; Nicholas Honerkamp, "Innovation and Change in the Antebellum Southern Iron Industry: An Example From Chattanooga, Tennessee," *IA* 13, no. 1 (1987): 55–68.

5. Plan view maps record the horizontal positions and extent of the archaeological remains; they provide a sort of bird's-eye view of what is found. Profile maps are records of the stratigraphic layers seen in the walls (or profiles) of each excavation unit dug.

6. This is an example of one of the most useful dating tools of archaeological fieldwork: the law of superposition. Simply stated, lower deposits *must* be earlier in time than any deposits that lie above or cut into them.

7. Brown had exposed this same feature in his 1977 excavations.

8. We were unable to identify the manufacturer of this singular example.

9. Sanders and Gould, *History Cast in Metal*, 27.

Chapter 5

1. White, "X-Ray Fluorescent Analysis of an Early Ohio Blast Furnace Slag," 186–88, and "Archaeological and Chemical Evidence for the Earliest American Use of Raw Coal as a Fuel in Ironmaking," 391–93. John R. White, "Analysis of the Lithic Raw Materials From the Earliest Blast Furnace West of the Alleghany Mountains: The Eaton (Hopewell) 1802–1808," manuscript, Youngstown State Univ., 1980; "The Eaton Blast Furnace," *Current Anthropology* 21 (1980): 513–14; "Historic Blast Furnace Slags: Archaeological and Metallurgical Analysis," *Journal of the Historical Metallurgy Society* 14 (1980): 55–64; "New Light on Early American Ironmaking: The Eaton Furnace," *Journal of the Historical Metallurgy Society* 15 (1981): 88–93; "Analysis and Evaluation of the Raw Materials Used in the Eaton (Hopewell) Furnace, *Ohio Journal of Science* 82 (1982): 23–27.

2. John R. White tackled the sampling aspect of blast-furnace archaeology in "Early Nineteenth Century Blast Furnace Charcoals: Analysis and Economics," *The Conference on Historic Site Archaeology Papers* (Columbia, S.C.: Institute of Archeology and Anthropology), 15 (1983): 106–22.

3. For a description of this material, see Council, Will, and Honerkamp, "Bluff Furnace," 111–18.

4. Our group-class-type format takes its inspiration from the pattern recognition approach used by Stanley South in *Method and Theory in Historical Archaeology* (New York: Academic Press, 1977).

5. This example is almost identical to the "typical" framing chisel illustrated by Eric Sloane in *A Museum of Early American Tools* (New York: Ballantine, 1964), 53.

6. Although this hook undoubtedly functioned to fasten items together, it could just as easily (and accurately) be placed in the tools group.

7. Found in a disturbed context in the steam-boiler area, this item could have been used in the later house that was built over the industrial remains. However, door hinges had long replaced pintles by the time the house was built at the turn of the century, so we attribute this artifact to the furnace period.

8. Johnson, *The Principles, Operation and Products of the Blast Furnace*, 460.

9. Ibid., 465.

10. Ibid., 465–56.

11. Analysis of the iron produced at the Eaton-Hopewell Furnace in Ohio gives an average sulfur content of 0.086 percent (White, "Archaeological and Chemical Evidence," 392); this is considerably higher than the values for the Bluff Furnace iron and is not surprising, considering the low desulfurization indices of the Eaton-Hopewell slags.

12. Johnson, *The Principles, Operation and Products of the Blast Furnace*, 469.

13. Ibid., 468.

14. Overman, *The Manufacture of Iron*, 467.

15. Chapman, *The Iron and Steel Industries of the South*, 34, notes the significance of the Rockwood Formation to Tennessee's post–Civil War iron industry.

16. Safford, *Geology of Tennessee*, 457.

17. Lesley, *The Iron Manufacturer's Guide*, 83; Burchard, *The Red Iron Ores of East Tennessee*, 74.

18. Fairbairn, *Iron*, 25, 36.

19. Burchard, *The Red Iron Ores of East Tennessee*, 77. The Bessemer process converts molten iron from the blast furnace directly into steel, but only low-phosphorus ores can be used.

20. Johnson, *The Principles, Operation and Products of the Blast Furnace*, 169–70.

21. Fairbairn, *Iron*, 61.

22. White, "Historic Blast Furnace Slags," 57.

23. White, "X-Ray Fluorescent Analysis," 186–87.

24. Johnson, *The Principles, Operation and Products of the Blast Furnace*, 200–210.

25. White, "X-Ray Fluorescent Analysis," 188.

Chapter 6

1. Charles B. Dew, *Ironmaker to the Confederacy: Joseph R. Anderson and the Tredegar Iron Works* (New Haven, Conn.: Yale Univ. Press, 1966), 332.

2. An early treatment is by Lester J. Cappon, "History of the Southern Iron Industry to the Close of the Civil War," master's thesis, Harvard Univ., 1928. A modern regional treatment is Smith, "Historical Geography of the Southern Charcoal Iron Industry." Smith analyzes iron production by smelting regions, making the distinction of state boundaries superfluous. Chattanooga was classified as being in the Upland South iron-smelting region, legitimately associating it with furnaces in northern Georgia and western Virginia.

3. A regional archaeological site survey, not involving intensive site excavation, is exampled by Smith, Stripling, and Brannon, *A Cultural Resource Survey of Tennessee's Western Highland Rim Iron Industry, 1790s–1930s*. A detailed recording, and historical survey of, a Civil War–period furnace that was never put into blast is provided by Joseph and Reed, "Ore, Water, Stone and Wood: Historical and Architectural Investigations of Donaldson's Iron Furnace, Cherokee County, Georgia." An extensive excavation program at a unique antebellum iron smelter in Alabama is reported by DeJarnette and DeJarnette, "Geology and Archaeology of the Old Tannehill Furnace Site."

4. Seely, "Blast Furnace Technology," 27–54.

5. See Robert C. Allen, "The Peculiar Productivity History of American Blast Furnaces, 1840–1913," *Journal of Economic History* 37 (1977): 605–33.

6. Seely, "Blast Furnace Technology," 35.

7. Ibid., 47.

8. Laurel Furnace on the Western Highland Rim, abandoned in 1856, produced pig iron at a rate of 16.4 tons per week. This is the smallest per-week production figure of any Middle Tennessee furnace calculable from the Lesley data.

9. Swank, *History of the Manufacture of Iron*, 370.

10. An excellent treatment of coke-related changes in iron production technology and economics is Hyde, *Technological Change and the British Iron Industry, 1700–1870*.

11. Dimensions given by Armes, *The Story of Coal and Iron in Alabama*, 175.

12. The operations of the East Tennessee Iron Manufacturing Company's Roane County furnace and forge are virtually unknown. It is conceivable that Eagle Furnace and Forge was the economic mainstay of the company. The specific allocation of any development capital after 1852 is unknown.

13. Honerkamp, "Innovation and Change in the Antebellum Southern Iron Industry," 55–68.

14. Whiteside was apparently a resident of Nashville from 1857 to 1860 according to Gaston, "The Remarkable Harriet Whiteside," 337.

15. Thomas C. Cochran, "Did the Civil War Retard Industrialization?" *Mississippi Valley Historical Review* 48 (1961): 210.

16. From data in Lesley, *The Iron Manufacturer's Guide*.

17. For example, South Carolina's iron production had peaked in the 1840s and 1850s; the use of inferior-grade ores, the lack of railroad transportation, and depletion of timberland around the furnaces left iron production in the state on its last legs when the war erupted and temporarily infused life into the industry. Ernest M. Lander, Jr., "The Iron Industry in Ante-Bellum South Carolina," *Journal of Southern History* 20 (1954): 337–55.

18. Swank, *History of the Manufacture of Iron*, 292.

19. A comprehensive study of Chattanooga's iron manufacturing industries has yet to be produced. General summaries appear in the several city histories, notably in Govan and Livingood, *The Chattanooga Country*, and Wilson, *Chattanooga's Story*.

20. Killebrew, *Introduction to the Resources of Tennessee*, 526.

21. *The Daily Times*, Chattanooga, July 29, 1882.

22. Morrow Chamberlain, *A Brief History of the Pig Iron Industry of East Tennessee* (Chattanooga: pub. by author, 1942), 16.

23. Ibid., 20–21. The mining properties were worked until 1918, when they were sold to the Alabama Company (later merged with Sloss-Sheffield Coal and Iron Company, Birmingham). The remaining assets of the company were liquidated in 1939.

24. Swank, *History of the Manufacture of Iron*, 291.

25. James M. Swank, *Classified List of Rail Mills and Blast Furnaces in the United States* (Philadelphia: American Iron and Steel Association, 1873), 28–30.

26. Killebrew, *Introduction to the Resources of Tennessee*, 227.

27. *The Daily Times*, Chattanooga, Jan. 23, 1878.

28. Doster, "The Chattanooga Rolling Mill," 45–55.

29. Chamberlain, *A Brief History of the Pig Iron Industry of East Tennessee*, 1942.

30. Constantine Belissary, "The Rise of Industry and the Industrial Spirit in Tennessee, 1865–1885," *Journal of Southern History* 19 (1953): 193–215. Coal and iron-ore mining in the state increased steadily during the late nineteenth century, a growth made possible by the construction of mining railroads.

31. Livingood, *A History of Hamilton County, Tennessee*, 57.

32. Chapman, *The Iron and Steel Industries of the South*, 103–4.

33. Joel A. Tarr, "The Separate vs. Combined Sewer Problem: A Case Study in Urban Technology Design Choice," *Journal of Urban History* 5 (1979): 308–39.

34. This was the Machine, Foundry and Pipe Works of D. Giles and Company, Henry and Sidney streets, Chattanooga.

35. Chapman, *The Iron and Steel Industries of the South*, 104. The consolidation of these pipe-founding plants is a good example of horizontal integration.

36. Discussed in Smith, "Historical Geography of the Southern Charcoal Iron Industry," 38.

37. Constantine G. Belissary, "Industry and Industrial Philosophy in Tennessee, 1850–1860," *East Tennessee Historical Society's Publications* 23 (1951): 46–57.

38. For a Deep South example, see Robert S. Davis, Jr., "Robert Findlay: Antebellum Iron Founder of Macon," *Journal of Southwest Georgia History* 3 (1985): 26–43.

39. Fred Bateman, James Foust, and Thomas Weiss, "The Participation of Planters in Manufacturing in the Antebellum South," *Agricultural History* 48 (1974): 289.

40. Susanna Delfino, "Antebellum East Tennessee Elites and Industrialization: The Examples of the Iron Industry and Internal Improvements," *East Tennessee Historical Society's Publications* 56–57 (1984–85): 102–19.

41. A comprehensive treatment of the economic relationship between slavery and industrialization is presented by Eugene D. Genovese, *The Political Economy of Slavery: Studies in the Economy and Society of the Slave South* (New York: Random House, 1965).

42. Fred Bateman and Thomas Weiss, *A Deplorable Scarcity: The Failure of Industrialization in the Slave Economy* (Chapel Hill: Univ. of North Carolina Press, 1981), 30–34.

43. See R. Keith Aufhauser, "Slavery and Technological Change," *Journal of Economic History* 34 (1974): 36–50.

44. "The iron industry's continued reliance upon Negroes in the 1840s and 1850s probably contributed to the failure of ironmasters to adopt improved methods of production, and thus put them at an increasing disadvantage with regard to Northern and British iron in those decades. In general, slave labor brought neither satisfaction nor progress to Virginia's ironworkers and their industry." S. Sydney Bradford, "The Negro Ironworker in Ante Bellum Virginia," *Journal of Southern History* 25 (1959): 206.

45. Robert S. Starobin, *Industrial Slavery in the Old South*, 149. The money invested in slaves was lost at the end the Civil War, of course, hampering the recovery of southern iron works during Reconstruction. Slaves and land were the bulk of prewar capital in the region.

46. Fred Bateman, James Foust, and Thomas Weiss, "Profitability in Southern Manufacturing: Estimates for 1860," *Explorations in Economic History* 12 (1975): 211–31.

47. Fred Bateman and Thomas Weiss, "Comparative Regional Development in Antebellum Manufacturing," *Journal of Economic History* 35 (1975): 185; Jeremy Atack, "Returns to Scale in Antebellum United States Manufacturing," *Explorations in Economic History* 14 (1977): 337–59.

48. Mark P. Leone, "Archaeology as the Science of Technology: Mormon Town Plans and Fences," in *Research and Theory in Current Archaeology*, ed. Charles L. Redman (New York: John Wiley and Sons, 1973), 125–50.

49. This is an old epistemological issue that we will not go into here. What constitutes "historical accuracy" versus "confirmation of the status quo" is not always intuitively obvious.

Bibliography

Published Sources

Ainsworth, Brig. Gen. Fred C., and Joseph W. Kirkley, comps. 1900. *The War of the Rebellion: A Compilation of the Official Records of the Union and Confederate Armies*. Ser. 4, vol. 3. Washington, D.C.: Government Printing Office.

Allen, Penelope Johnson. 1934. "Leaves from the Family Tree: Cravens." *Chattanooga Times*, Mar. 18.

Allen, Robert C. 1977. "The Peculiar Productivity History of American Blast Furnaces, 1840–1913." *Journal of Economic History* 37: 605–33.

American Iron and Steel Association. 1880. *Directory to the Iron and Steel Works of the United States*. Philadelphia.

Armes, Ethel. 1910. *The Story of Coal and Iron in Alabama*. Birmingham, Ala.: Birmingham Chamber of Commerce. Rpt. 1973. New York: Arno Press.

Atack, Jeremy. 1977. "Returns to Scale in Antebellum United States Manufacturing." *Explorations in Economic History* 14: 337–59.

Aufhauser, R. Keith. 1974. "Slavery and Technological Change." *Journal of Economic History* 34: 36–50.

Bateman, Fred, James Foust, and Thomas Weiss. 1974. "The Participation of Planters in Manufacturing in the Antebellum South." *Agricultural History* 48: 277–97.

———. 1975. "Profitability in Southern Manufacturing: Estimates for 1860." *Explorations in Economic History* 12: 211–31.

Bateman, Fred, and Thomas Weiss. 1975. "Comparative Regional Development in Antebellum Manufacturing." *Journal of Economic History* 35: 182–208.

———. 1981. *A Deplorable Scarcity: The Failure of Industrialization in the Slave Economy*. Chapel Hill: Univ. of North Carolina Press.

Bayley, W. S. 1923. *The Magnetic Iron Ores of East Tennessee and Western North Carolina*. Nashville: Tennessee Dept. of Education, Div. of Geology, Bulletin 29.

Belissary, Constantine G. 1951. "Industry and Industrial Philosophy in Tennessee, 1850–1860." *East Tennessee Historical Society's Publications* 23: 46–57.

———. 1953. "The Rise of Industry and the Industrial Spirit in Tennessee, 1865–1885." *Journal of Southern History* 19: 193–215.

Bennett, James R. 1986. *Old Tannehill: A History of the Pioneer Ironworks in Roupes Valley, 1829–1865*. Birmingham, Ala.: Jefferson County Historical Commission.

Binford, Lewis R. 1962. "Archaeology as Anthropology." *American Antiquity* 28, no. 2: 217–25.

Boney, F. N. 1977. "Part Three: 1820–1865." In *A History of Georgia*, ed. Kenneth Coleman. Athens: Univ. of Georgia Press, 127–204.

Bradford, S. Sydney. 1959. "The Negro Ironworker in Ante Bellum Virginia." *Journal of Southern History* 25: 194–206.

Brown, Jeffrey L. 1977. *Exploratory Archaeological Excavations at the Bluff Furnace Site.* Miscellaneous Paper 1, Institute of Archaeology, Univ. of Tennessee at Chattanooga.

———. 1978. "The Camp Site Beneath the Cravens House Porch." *Tennessee Anthropologist* 3: 6–13.

Bryant, William Cullen, ed. 1872. *Picturesque America*, vol. 1. D. New York: Appleton and Co.

Burchard, Ernest F. 1913. *The Red Iron Ores of East Tennessee.* Nashville: Tennessee State Geological Survey, Bulletin 16. Rpt. 1962.

———. 1934. *The Brown Iron Ores of the Western Highland Rim, Tennessee.* Nashville: Tennessee Div. of Geology, Bulletin 39.

Caldwell, Joshua W.1898. *Sketches of the Bench and Bar of Tennessee.* Knoxville: Ogden Brothers and Co.

Campbell, T. J. 1932. *The Upper Tennessee: Comprehending Desultory Records of River Operations in the Tennessee Valley.* Chattanooga: pub. by author.

Chamberlain, Morrow. 1942. *A Brief History of the Pig Iron Industry of East Tennessee.* Chattanooga: pub. by author.

Chapman, Herman H. 1953. *The Iron and Steel Industries of the South.* Birmingham: Univ. of Alabama.

Clark, Victor S. 1949. *History of Manufactures in the United States, Volume II 1860–1893.* New York: Peter Smith. Rpt. of 1929 ed.

Cochran, Thomas C. 1961. "Did the Civil War Retard Industrialization?" *Mississippi Valley Historical Review* 48: 197–210.

Crocker, Les. 1973. "An Early Iron Foundry in Northern Mississippi." *Journal of Mississippi History* 35: 113–27.

Dalton, Robert E. 1976. "Montgomery Bell and the Narrows of Harpeth." *Tennessee Historical Quarterly* 35: 3–28.

Davis, Major George B., Leslie J. Perry, and Joseph W. Kirkley, compilers. 1890. *The War of the Rebellion, A Compilation of the Official Records of the Union and Confederate Armies.* Ser. 1, vol. 31. Washington, D.C.: Government Printing Office.

Davis, Robert S., Jr. 1985. "Robert Findlay: Antebellum Iron Founder of Macon." *Journal of Southwest Georgia History* 3: 26–43.

Deaderick, Lucille, ed. 1976. *Heart of the Valley: A History of Knoxville, Tennessee.* Knoxville: East Tennessee Historical Society.

DeBow, J. D. B. 1854. *A Statistical View of the United States . . . Being a Compendium of the Seventh Census.* Washington, D.C.: A. O. P. Nicholson, Public Printer.

Delfino, Susanna. 1984–85. "Antebellum East Tennessee Elites and Industrialization: The Examples of the Iron Industry and Internal Improvements." *East Tennessee Historical Society's Publications* 56–57: 102–19.

Dew, Charles B. 1966. *Ironmaker to the Confederacy: Joseph R. Anderson and the Tredegar Iron Works.* New Haven, Conn.: Yale Univ. Press.

———. 1975. "Black Ironworkers and the Slave Insurrection Panic of 1856." *Journal of Southern History* 41: 321–38.

Doster, James F. 1964. "The Chattanooga Rolling Mill: An Industrial By-Product of the Civil War." *East Tennessee Historical Society's Publications* 36: 45–55.

Fagan, Brian. 1986. *People of the Earth.* Boston: Little, Brown.

Fairbairn, William. 1865. *Iron; Its History, Properties and Processes of Manufacture.* Edinburgh: Adam and Charles Black.

Fritz, John. 1912. *The Autobiography of John Fritz.* New York: John Wiley and Sons.

Gaston, Kay Baker. 1981. "The Remarkable Harriet Whiteside." *Tennessee Historical Quarterly* 40: 333–47.

Genovese, Eugene D. 1965. *The Political Economy of Slavery: Studies in the Economy and Society of the Slave South.* New York: Random House.

Goff, Richard D. 1969. *Confederate Supply.* Durham, N.C.: Duke Univ. Press.

Gordon, Robert B. 1983. "Materials for Manufacturing: The Response of the Connecticut Iron Industry to Technological Change and Limited Resources." *Technology and Culture* 24: 602–34.

Govan, Gilbert E., and James W. Livingood. 1977. *The Chattanooga Country, 1540–1976: From Tomahawks to TVA.* Third ed., rev. and updated by James W. Livingood. Knoxville: Univ. of Tennessee Press.

Guernsey, Alfred H., and Henry M. Alden. 1868. *Harper's Pictorial History of the Civil War.* Vol. 2. Chicago: McDonell Brothers.

Hale, Will T., and Dixon L. Merritt. 1913. *A History of Tennessee and Tennesseans.* Vol. 6. New York: Lewis Publishing Co.

Hartley, E. N. 1957. *Ironworks on the Saugus.* Norman: Univ. of Oklahoma Press.

Hershey, Robert E., and Stuart W. Maher. 1963. *Limestone and Dolomite Resources of Tennessee.* Nashville: Tennessee Div. of Geology, Bulletin 65.

Holbrook, Stewart H. 1939. *Iron Brew: A Century of American Ore and Steel.* New York: Macmillan Co.

Honerkamp, Nicholas. 1987. "Innovation and Change in the Antebellum Southern Iron Industry: An Example From Chattanooga, Tennessee." *IA* 13, no. 1: 55–68.

Hudson, Kenneth. 1979. *World Industrial Archaeology.* Cambridge, England: Cambridge Univ. Press.

Hunt, Raymond F., Jr. 1966. "The Pactolus Ironworks." *Tennessee Historical Quarterly* 25: 176–96.
Hyde, Charles K. 1977. *Technological Change and the British Iron Industry, 1700–1870.* Princeton, N.J.: Princeton Univ. Press.
Johnson, J. E., Jr. 1918. *The Principles, Operation and Products of the Blast Furnace.* New York: McGraw-Hill.
Kesey, Ken. 1977 *Sometimes a Great Notion.* New York: Penguin Books.
Killebrew, J. B. 1874. *Introduction to the Resources of Tennessee: First and Second Reports of the Bureau of Agriculture, State of Tennessee.* Nashville: Tavel, Eastman and Howell, State Printers.
———. 1881. *Iron and Coal in Tennessee: Report by the Commissioner of Agriculture, Statistics and Mines.* Nashville: Tennessee Bureau of Agriculture.
Lander, Ernest M., Jr. 1954. "The Iron Industry in Ante-Bellum South Carolina." *Journal of Southern History* 20: 337–55.
———. 1960. "Charleston: Manufacturing Center of the Old South." *Journal of Southern History* 26: 330–51.
Lane, Peter. 1978. *The Industrial Revolution: The Birth of the Modern Age.* New York: Barnes and Noble.
Leone, Mark P. 1973. "Archaeology as the Science of Technology: Morman Town Plans and Fences" In *Research and Theory in Current Archaeology*, ed. Charles L. Redman. New York: John Wiley and Sons. 125–50.
Lesley, J. Peter. 1859. *The Iron Manufacturer's Guide to the Furnaces, Forges and Rolling Mills of the United States.* New York: J. Wiley.
Lewis, J. Eugene. 1961. "Cravens House, Landmark of Lookout Mountain." *Tennessee Historical Quarterly* 20: 203–21.
Lewis, Ronald L. 1979. *Coal, Iron, and Slaves: Industrial Slavery in Maryland and Virginia, 1715–1865.* Contributions in Labor History, Number 6. Westport, Conn.: Greenwood Press.
Lewis, W. David. 1976. *Iron and Steel in America.* Greenville, Del.: Hagley Museum.
Livingood, James W. 1947. "Chattanooga: A Rail Junction of the Old South." *Tennessee Historical Quarterly* 6: 230–50.
———. 1981. *A History of Hamilton County, Tennessee.* Memphis: Memphis State Univ. Press.
Long, John P. 1880. "Early Days of Chattanooga." In *Transactions of the Iron, Coal and Manufacturer's Association of Chattanooga, Tennessee, for the Year 1880*, 15–18. Chattanooga: Iron, Coal and Manufacturer's Association of Chattanooga.
Luther, Edward T. 1959. *The Coal Reserves of Tennessee.* Nashville: Tennessee Div. of Geology, Bulletin 63.
———. 1977. *Our Restless Earth: The Geologic Regions of Tennessee.* Knoxville: Univ. of Tennessee Press.

Maher, Stuart W. 1964. *The Brown Iron Ores of East Tennessee*. Nashville: Tennessee Dept. of Conservation, Div. of Geology, Report of Investigations, no. 19.

Malone, Dumas, ed. 1933. *Dictionary of American Biography* 10. New York: Charles Scribner's Sons.

———. 1936. *Dictionary of American Biography* 18. New York: Charles Scribner's Sons.

McGuffey, Charles D., ed. 1911. *Standard History of Chattanooga, Tennessee*. Knoxville: Crew and Dorsey.

McKenzie, Robert H. 1972. "Reconstruction of the Alabama Iron Industry, 1865–1880." *Alabama Review* 25: 178–91.

Mulholland, James A. 1981. *A History of Metals in Colonial America*. University: Univ. of Alabama Press.

Nelson, Wilbur A. 1925. *The Southern Tennessee Coal Field*. Nashville: Tennessee Dept. of Education, Div. of Geology, Bulletin 33-A.

Ormsbee, J. J. 1898. "The Rise and Progress of Coal Mining in Tennessee." In A. D. Hargis, *Seventh Annual Report of the Bureau of Labor, Statistics and Mines of the State of Tennessee, for the year ending December 31, 1897*. Nashville: Breeder and Horseman.

Overman, Frederick. 1854. *The Manufacture of Iron in All Its Various Branches*. Philadelphia: H. C. Baird.

Parham, Lou L. 1871. *Parham's First Annual Directory of the City of Chattanooga*. Knoxville: Whig and Register Steam Print.

Paskoff, Paul F. 1983. *Industrial Evolution: Organization, Structure, and Growth of the Pennsylvania Iron Industry, 1750–1860*. Baltimore: Johns Hopkins Univ. Press.

Pickel, Eugene Monroe. 1981. *A History of Roane County, Tennessee, to 1860*. Kingston, Tenn.: Roane County Heritage Commission.

Pollard, Helen P. 1987. "The Political Economy of Prehispanic Tarascan Metallurgy." *American Antiquity* 52, no. 4: 741–52.

Phillips, William Battle. 1912. *Iron Making in Alabama*. 3d ed. University: Geological Survey of Alabama.

Price, R. N. 1913. *Holston Methodism, From Its Origin to the Present Time*. Vol. 5. Nashville: Publishing House of the Methodist Episcopal Church, South.

Raulston, J. Leonard, and James W. Livingood. 1974. *Sequatchie: A Story of the Southern Cumberlands*. Knoxville: Univ. of Tennessee Press.

Rogers, John. 1948. *Geology and Mineral Deposits of Bumpass Cove, Unicoi and Washington Counties, Tennessee*. Nashville: Tennessee Dept. of Conservation, Div. of Geology, Bulletin 54.

Rutledge, John W., and Robert B. Gordon. 1987. "The Work of Metallurgical Artificers at Machu Picchu, Peru." *American Antiquity* 52, no. 3: 578–94.

Safford, James M. 1869. *Geology of Tennessee*. Nashville: S. C. Mercer.

———, and Joseph B. Killebrew. 1904. *The Elements of the Geology of Tennessee.* Nashville: Ambrose and Bostelman.

Sanders, Clyde A., and Dudley C. Gould. 1976. *History Cast in Metal: The Founders of North America.* Des Plaines, Ill.: Cast Metals Institute of the American Foundrymen's Society.

Schroeder, D. L., and K. C. Ruhl. 1968. "Metallurgical Characteristics of North American Prehistoric Copper Work." *American Antiquity* 33, no. 2 : 162–69.

Seely, Bruce E. 1981. "Blast Furnace Technology in the Mid-Nineteenth Century: A Case Study of the Adirondack Iron and Steel Company." *IA* 7, no. 1: 27–54.

Shallenberg, Richard. 1975. "Evolution, Adaptation, and Survival: The Very Slow Death of the American Charcoal Iron Industry." *Annals of Science* 32, no. 4: 341–58.

Shallenberg, Richard, and David A. Ault. 1977. "Raw Materials Supply and Technological Change in the American Charcoal Iron Industry." *Technology and Culture* 18: 436–66.

Sloane, Eric. 1964. *A Museum of Early American Tools.* New York: Ballantine.

Smith, Samuel D., Charles P. Stripling, and James M. Brannon. 1988. *A Cultural Resource Survey of Tennessee's Western Highland Rim Iron Industry, 1790s–1930s.* Research Series, no. 8. Nashville: Tennessee Dept. of Conservation, Div. of Archaeology.

South, Stanley. 1977. *Method and Theory in Historical Archaeology.* New York: Academic Press.

Starobin, Robert S. 1970. *Industrial Slavery in the Old South.* New York: Oxford Univ. Press.

Stoughton, Bradley. 1913. *The Metallurgy of Iron and Steel.* Second ed. New York: McGraw-Hill Book Co.

Strother, David Hunter. Engraving of Bluff Furnace. 1858. *Harpers New Monthly Magazine* 17: 298.

Swank, James M. 1873. *Classified List of Rail Mills and Blast Furnaces in the United States,* Philadelphia: American Iron and Steel Association.

———. 1878. *Introduction to a History of Ironmaking and Coal Mining in Pennsylvania.* Philadelphia: pub. by author.

———. 1892. *History of the Manufacture of Iron in All Ages, and Particularly in the United States from Colonial Times to 1891.* Second ed. Philadelphia: American Iron and Steel Association.

Tarr, Joel A. 1979. "The Separate vs. Combined Sewer Problem: A Case Study in Urban Technology Design Choice." *Journal of Urban History* 5: 308–39.

Tennessee, State of. 1848. *Acts of the State of Tennessee passed at the First Session of the Twenty-Seventh General Assembly, for the years 1847–8.* Jackson, Tenn.: Gates and Parker.

---. 1852. *Acts of the State of Tennessee passed at the First Session of the Twenty-Ninth General Assembly, for the years 1851–2.* Nashville: M'Kennie and Brown.

---. 1854. *Acts of the State of Tennessee passed at the First Session of the Thirtieth General Assembly, for the years 1853–4.* Nashville: M'Kennie and Brown.

TVA. 1959. *Floods on Tennessee River, Chattanooga and Dry Creeks, and Stringers Branch, Vicinity of Chattanooga, Tennessee.* Knoxville: TVA Div. of Water Control Planning.

Tylecote, R. F. 1976. *A History of Metallurgy.* London: Metals Society.

Ure, Andrew. 1868. *A Dictionary of Arts, Manufactures and Mines; Containing a Clear Exposition of their Principles and Practice.* Vol. 1. New York: D. Appleton and Co.

White, John R. 1977. "X-Ray Fluorescent Analysis of an Early Ohio Blast Furnace Slag." *Ohio Journal of Science* 77: 186–88.

---. 1978. "Archaeological and Chemical Evidence for the Earliest American Use of Raw Coal as a Fuel in Ironmaking." *Journal of Archaeological Science* 5: 391–93.

---. 1980. "The Eaton Blast Furnace." *Current Anthropology* 21: 513–14.

---. 1980. "Historic Blast Furnace Slags: Archaeological and Metallurgical Analysis." *Journal of the Historical Metallurgy Society* 14: 55–64.

---. 1981. "New Light on Early American Ironmaking: The Eaton Furnace." *Journal of the Historical Metallurgy Society* 15: 88–93.

---. 1982. "Analysis and Evaluation of the Raw Materials Used in the Eaton (Hopewell) Furnace." *Ohio Journal of Science* 82: 23–27.

---. 1983. "Early Nineteenth Century Blast Furnace Charcoals: Analysis and Economics." *The Conference on Historic Site Archeology Papers* (Columbia, S.C.: Institute of Archeology and Anthropology), 15: 106–22.

Wilson, John. 1980. *Chattanooga's Story.* Chattanooga: Chattanooga News–Free Press.

Wiltse, Henry M. 1916–19. *History of Chattanooga.* 2 vols. Chattanooga: pub. by author.

Wish, Harvey. 1939. "The Slave Insurrection Panic of 1856." *Journal of Southern History* 5: 206–22.

Woodward, Joseph H., II. 1940 *Alabama Blast Furnaces.* Woodward, Ala.: Woodward Iron Co.

---. 1954. "Alabama Iron Manufacturing, 1860–1865." *Alabama Review* 7: 199–207.

Unpublished Sources

Cappon, Lester J. 1928. "History of the Southern Iron Industry to the Close of the Civil War." Master's thesis. Harvard Univ.

Council, R. Bruce, M. Elizabeth Will, and Nicholas Honerkamp. 1982. "Bluff Furnace: Archaeology of a Nineteenth Century Blast Furnace." Manuscript. Jeffrey L. Brown Institute of Archaeology, Univ. of Tennessee at Chattanooga.

DeJarnette, David, and Thomas H. DeJarnette, Jr. 1956. "Geology and Archaeology of the Old Tannehill Furnace Site." Report. Tannehill Historical State Park, McCalla, Ala.

Dorr, F. W. 1863. *Chattanooga and Its Approaches, showing the Union and Rebel Works before and during the Battles 23rd, 24th and 25th November, 1863. Surveyed under the direction of Brig. Gen. Wm. F. Smith, Chief Engineer of the Military Division of the Mississippi, during parts of November and December 1863.* U.S. Coast Survey. Original in the Library of Congress. Chattanooga: TVA Mapping Services Div.

Evans, E. Raymond, and Vicky Karhu. 1985. "Cultural Overview and Synthesis Study of the Chattanooga Riverfront, Chattanooga, Tennessee." Report prepared for Moccasin Bend Task Force, City of Chattanooga and Hamilton County.

Fritz, John. Correspondence. Canal Museum, Hugh Moore Historical Park and Museums. Easton, Pa.

Hamilton County, Tenn. Hamilton County Deed Books. Hamilton County Courthouse, Chattanooga.

Joseph, J. W., and Mary Beth Reed. 1988. "Ore, Water, Stone and Wood: Historical and Architectural Investigations of Donaldson's Iron Furnace, Cherokee County, Georgia." Report. Atlanta: Garrow and Associates.

Marion County, Tenn. Marion County Deed Books. Marion County Courthouse, Jasper.

Roane County, Tenn. Roane County Deed Books. Roane County Courthouse, Kingston.

Rolando, Victor R. 1980. "Ironmaking in Vermont: 1775–1890." Master's thesis. College of Saint Rose, Albany, N.Y.

Smith, James Larry. 1982. "Historical Geography of the Southern Charcoal Iron Industry, 1800–1860." Ph.D. thesis. Univ. of Tennessee, Knoxville.

Synder, Mary, and Phil Noblitt. 1975. "Robert Cravens: A Brief History of the Man and His Family." Manuscript. Chickamauga and Chattanooga National Military Park.

U.S. Census Bureau. 1860. *Eighth Census of the United States, 1860.* Work Sheets. Population Schedule 1. Free Inhabitants, Hamilton County, Tenn. Microfilm.

White, John R. 1980. "Analysis of the Lithic Raw Materials From the Earliest Blast Furnace West of the Alleghany Mountains: The Eaton (Hopewell) 1802–1808." Manuscript. Dept. of Sociology/Anthropology, Youngstown State Univ.

Newspapers

Chattanooga *Daily Rebel.*
Chattanooga *Gazette.*
Chattanooga *News.*
Chattanooga *Times.*
Daily American Union (Knoxville).

Index

Adirondack Iron and Steel Company (New York), 9
Agnew, Samuel J., 67
agrarianism of prewar South, 175
Alley, Erasmus, 66, 201n.56
American Pipe and Foundry Company, 174
anchor bolts, at Bluff Furnace, 102-4, 103 (illus.)
Andreae, Edward E., 49
anthracite coal: application of hot blast to, 17-18, 33; defined, 15
archaeology, historical, 1
archaeology, industrial: defined, 1, 2; examples of, 1, 2, 9; objectives of, 2-3
Armes, Ethel, 76
artifacts, from Bluff Furnace. *See* Bluff Furnace (artifacts)

Baltimore Company, 193n.11
Bell, Montgomery, 46, 56
bell hopper, 33
beneficiation of iron ore, 13
Bessemer steel, 171, 208n.19
Bethlehem Iron Company, 74
Bethlehem Iron Works, 83
Bethlehem Steel, 74
Birmingham, Ala.: early steel manufacture at, 87; postwar iron production of, 174
bituminous coal: defined, 15; mentioned, 17, 18, 45, 63, 170
blast (air): defined, 26; delivery of, 29; preheating of, 196n.20; volumes and pressures, 37
blast bellows. *See* blast machines
blast failures, 40-41; *see also* scaffolds, salamander
blast furnace: chemistry of smelting in, 35-37; components of, 20-35; general configuration of, 20-21; management of operations, 37; operations described, 11; temperatures reached in, 36-37; work routine at, 40; *see also* bell hopper, blast machines, blast nozzle, bosh, bustle pipe, casting shed, charging deck, cinder notch, downcomer pipe, forebay, hearth, mantle, skip-hoist, stack, tunnel head, tuyeres
blast furnace fuels: economics of use, 17; historical trends in U.S., 18-19
blast furnace technology, historical research concerning, 11
blast machines: blast bellows, 27; choice of motive power, 26-27; cylinder blast bellows, 27, 28 (illus.); cylinder blast pump, 30 (illus.); general requirements, 26; rotary blast fan, 29, 31 (illus.)
blast nozzle, 30
bloom, 20, 42
bloomary forges: described, 20; in East Tennessee, 46
blowing in, 37
Bluff Furnace (archaeology): activity areas defined, 97; blast machine foundations, 121, 124-25; casting shed excavations, 96 (illus.), 97, 107-12, 111 (illus.), 127; casting shed features, 96 (illus.), 98, 110, 111 (illus.), 112, 127; charcoal-fired limestone stack, 98, 113, 121, 122, 124, 129; charging deck excavations, 101-4; charging deck features 97, 102-4, 103 (illus.); coke-fired iron cupola 98, 114 (illus.), 115, 117 (illus.), 118, 120, 121, 122, 124, 129, 133; furnace cooling features, 121-23; furnace elements, 97, 115 (illus.), 116 (illus.), 117-18, 119, 120, 121, 122 (illus.), 123, 129, 130; furnace excavation, 100 (illus.), 112-26; furnace failure, 119, 120; pig iron

Bluff Furnace (archaeology) (*cont.*)
  from, 123, 137, 138, 144, 145 (illus.);
  pipe chase, 125 (illus.), 126; plan of
  excavations, 98; reburial and site
  stabilization, 128; reconstruction of
  plant, 128-30; salamander, 115, 118,
  119, 122; scale of, 127; site discovery,
  92; site excavation, 9-10, 92-127, 96
  (illus.),100 (illus.); steam boiler, 97,
  104-7, 105 (illus.), 122; threats to, 95-
  96; tuyeres, 115 (illus.), 117, 118, 122
Bluff Furnace (artifacts): analysis methods,
  133; classification format, 135, 136-38;
  coal, 137, 152, 153, 155; coke, 137,
  152, 154, 155; coke associated with
  salamander, 114 (illus.), 116; coke-
  versus charcoal-produced pig iron, 149,
  150; conservation of, 134, 135, domestic
  types, 135, 138, 146-47; fasteners, 136,
  139, 140 (illus.); furnace area examples,
  113 (illus.), 113-15, 123; industrial
  types, 131-59; iron-ore analysis, 137,
  145, 150-52, 159; miscellaneous
  artifacts 137, 142 (illus.), 143, 144
  (illus.); pig iron, 145 (illus.); pig-iron
  analysis, 148-50, 159, 160; pig-iron
  samples, 133, 137, 138, 144, 145, 148-
  50, 158, 159; sampling issues, 132; slag
  97, 126-27, 131, 137, 146 (illus.), 155-
  58; wrought iron tools, 136, 138-39,
  139 (illus.)
Bluff Furnace (history): abandonment and
  burial of, 6-7; archaeological site
  formation, 89-91; blast failure (1860),
  78; coal sources, 152, 159; charcoal
  operations commence, 62; coke iron
  production, 65, 73-74, 77; coke
  sources, 152, 159; commercial scope of,
  162; conversion from charcoal to coke,
  69-70, 72-73; demolition (1864), 83;
  derivation of name, 6; discovery of, 7;
  dismantlement (1862), 79-80; early
  production statistics, 62-63, 164;
  economic and technological evaluation,
  162-68; economy of scale at, 164-65;
  educational and ideological functions
  of, 177-78; engraving (1858), 64
  (illus.); furnace description in Lesley
  (1859), 63; history summarized, 161-
  62; iron ore sources, 150; lime kiln, 89,
  206n.129; location of, 6; mentioned,
  58, 64; photograph (1860), 73 (illus.),
  110, 111; photograph (1864), 84
  (illus.); photograph (1905), 90 (illus.),
  91; significance of, 6; site selection, 61;
  slave labor at, 76; workforce at, 75
Bluff View, 7, 90, 90 (illus.), 91
bosh or boshes, 21, 196n.13
Boyce, James Pettigru, 55, 66, 67, 80, 81
Boyce, Ker, 55, 65-66, 80, 89, 200n.31
Boyd, Samuel B., 53, 54
Brierfield Iron Works, 88
Bright Hope Furnace, 48
Bronze Age, 4
Brown, Dr, Jeffrey L., 7, 9, 11, 92, 95,
  194n.13
burden, 12
bustle pipe, 29

Cahawba Iron Works, 86
Cambria Iron Company, 75
Cambria Iron Works, 74, 75, 87
carbon absorption, 36
Carter, Farish, 54, 80
cast iron, 20, 41
cast-iron pipe production, 174
casting shed, 20-21, 35
Chamberlain, Hiram, 171
charcoal, 15
charcoal blast furnace, 22 (illus.); *see also*
  Bluff Furnace (archaeology): charcoal-
  fired limestone stack
charcoal-iron manufacture: at Bluff
  Furnace, 62; decline of, 196n.12;
  economics of, 18-19; southern commitment to, 19; Tennessee, 55-56;
  timberland reserves for, 16-17
charcoal mound, 16, 17 (illus.)
charcoaling: described, 15, 16; in mounds,
  16; in retorts, 16

charges: description of, 12; measurement of, 38
charging, 33-34
charging deck, 20, 33
Chattanooga Coal and Iron Company, 169
Chattanooga Creek industrial tract, 80, 62, 166
Chattanooga Foundry and Machine Shop, 68 (illus.); operations discussed, 68-69
Chattanooga Furnace, 169
Chattanooga Iron Company, 169
Chattanooga Iron Manufacturing Company, 200n.31
Chattanooga Island, 61
Chattanooga Leather Manufacturing Company, 89
Chattanooga, Tenn.: engraving (1863), 82 (illus.); mentioned, 5, 6, 7, 44, 58; military occupation of, 1864-65, 81-83; plat of, 59; postwar iron industry of, 168-74; railroads servicing, 50-52; steel manufacture at, 171
Chattanooga Waterworks Company, 55
cinder notch, 38
Citico Furnace, 169-70
Citico Furnace Company, 169-70
coal, anthracite. *See* anthracite coal
coal, bituminous. *See* bituminous coal
coal, Bluff Furnace: analysis of, 137, 152, 153, 155; sources of, 152, 158
coke: Bluff Furnace sources, 152, 158; characteristics of, in blast furnaces,18; defined,15; iron production using, 18-19; *see also* coke-iron manufacture
coke-iron manufacture: at Bluff Furnace, 65, 73-74, 77; at Eagle Furnace, 49; early, at Coalbrookdale, England, 198n.15; in South during Civil War; 86-87; in U.S., 18-19
coke oven, described, 18, 19 (illus.)
coking: discussed, 15, 18; of southern bituminous coals, 201n.54
colliers, 16
Confederate iron production, 85-87
context: archaeological, 132; historical, 10, 43

Cooper, Mark Anthony, 86
Cravens, Robert: at Bluff Furnace, 58, 63, 70, 71, 75; death of, 89; early life, 47-48; mentioned, 52, 53, 81, 90, 161, 162; portrait of, 52 (illus.), postwar activities of, 88-89; role in East Tennessee Iron Manufacturing Company, 54, 81; role in Etna Mines, 65-66, 159
crucible, furnace, 21; *see also* hearth
crucible steel, 42
Cumberland Furnace, 46
Cumberland Plateau, 44, 45, 55
cupola, foundry, 24
cupola blast furnace: described, 23-24; 25 (illus.), 26 (illus.); operation critiqued, 24; *see also* Bluff Furnace (archaeology): coke-fired iron cupola
cylinder blast pump. *See* blast machines

damstone, 21
Darby, Abraham, 198n.15
Dearing, Gilbert W., 62
Dearing Mill place, 62, 89
desulfurization index: defined, 157; estimates for Bluff Furnace slags, 156, 157; estimates for contemporary sites, 190
Dew, Charles B., 160
direct casting, 39-40
downcomer pipe, 32
Dyestone Belt: defined, 44; economic promise of, 45; mentioned, 174; postwar iron production in, 170; *see also* Tennessee: iron-producing regions

Eagle Furnace, 48-49, 54, 62-63, 64, 152, 163, 164, 201n.40
Eagle Furnace No. 2, 48
Earle, Colonel Elias, 47
East Tennessee Iron Manufacturing Company: charter granted, 53; division of assets, 1863, 80-81; Eagle Furnace operations of, 210n.12; early operations of, 54; foundry operations in Chatta-

East Tenn. Iron Mfg. Co. (*cont.*)
nooga, 58, 68; level of capitalization, 165; management structure of, 53-54, 167; mentioned, 49, 52; postwar dissolution of, 89-90; recapitalization in 1852, 61; summary of, 161, 162; vertical integration of, 165-67
East Tennessee: iron market restrictions, 50; physiography, 44-45; postwar iron manufacture, 174
East Tennessee, Virginia and Georgia Railroad, 52
Eastern Iron Belt: defined, 44; postwar iron production in, 170; *see also* Tennessee: iron-producing regions
Eastman, John W., 68
Eastman, Lees and Company, 68-69
Eaton Hopewell Furnace, 131, 132, 208n.11; analysis of slags from, 190-91
economy of scale: at blast furnaces, 24; limitations to, in East Tennessee iron production, 50; in prewar southern industry, 177
Edwards, Giles: at Bluff Furnace, 69-71, 75, 161; early career, 70; mentioned, 162; postwar activities, 87-88; at Shelby Iron Works, 79, 87-88
electrolysis, 134
Embree, Elihu and Elijah, 46
English, Mathew, 48
Etna Mines: economic link with Bluff Furnace, 166; incorporation, 66; mentioned, 63, 72, 152, 158; operations during Civil War, 84; production history, 65-67; production of coke at, 77-78
Etowah Iron Works, 86

features, definition of, 99
field techniques, archaeological, 95, 99
fillers, 33
firebricks, 21
fluxes: function of, 19; mentioned, 11, 12; types described, 20
Folger, Lafayette, 80

forebay, 21
foreman. *See* founder
Fort, Tomlinson, 90
founder, 38
foundry: of East Tennessee Iron Manufacturing Company, 58, 68-69, 166; operations described, 42
foundryman, 39
Fritz, John, 70, 71, 74, 83, 171
Fritz, William, 83
fuels, blast furnace: 15; *see also* charcoal, coal, coke

gangue, 12
gates, casting, 40, 143
Gordon, George, 48
Great Valley of East Tennessee, 44, 50, 58
guttermen, 39

Hamilton County, Tenn.: formation of, 47; industrial workers of, in 1860 Census, 188-89
hard-driving, 163
hearth, components of: damstone, 21; section through, 23 (illus.); timpstone, 21; *see also* crucible
hematite: described, 12; in the Dyestone Belt of Tennessee, 44, 174; in the Eastern Iron Belt of Tennessee, 44, 170; "fossil dyestone" iron ore used at Bluff Furnace, 150, 151, 152; used at Citico Furnace, 169
Henderson, James: at Bluff Furnace, 69-71, 75, 79, 83, 87, 166-68; patents of, 205n.118; postwar activities of, 161, 162
Henderson, Jane, 61
Henderson Steel and Manufacturing Company, 87
Hewitt, Abram S., 70, 74
Historic American Building Record, 9
Hittites, 4
Holly Springs Ironworks (Mississippi), 8
Hopewell Furnace (Ohio), 9

hot-blast stove: at Bluff Furnace, 130; described, 32, 34 (illus.); Neilson, 18
hot-blast technology: application to anthracite coal, 17-18; at Bluff Furnace, 64-65, 72-73, 163; characteristics of, 31-32; at Eagle Furnace, 49, 78; economics of, 33; temperatures at mid-19th century, 32

incipient fusion, 36
industrial railways at furnaces, 33
Industrial Revolution, 4
industrialization in the antebellum South, 175-77
innovation, technological: in antebellum southern industry, 176-77; at Bluff Furnace, 167-68
International Association of Metallurgists and Mineralogists, 87
inwall or inwalls, 21
iron manufacture: advantages of, 4; in antebellum South Carolina, 210n.17; bibliographic notes, 194-95n.1; history of, 3-5; lack of historical studies on, in the South, 160; New World introduction, 5; in the South, 5, 9-10; *see also* charcoal-iron manufacture, coke-iron manufacture
iron ore: Bluff Furnace sources of, 150; characteristics of, 12, 150; cleaning screen, 14 (illus.), composition of, 13; hematite, 12, 150; limonite, 12; magnetite, 13; preparations for use of, 13, 15; roasting in mounds, described, 13, 14 (illus.); Rock-wood Formation, 150; self-fluxing, 13
iron plantations, 41, 72, 166
ironmaster, 196n.24

J. Edgar Thomson Works, 87
James, J. J., 61, 62, 80
Johnson, J. E., 35, 148, 150
Johnson, Samuel, 55, 200n.28
Jones, William Richard ("Captain Bill"), 75-76, 87, 203n.85

keeper, 33
King, Walter, 46

ladlemen, 39
Lake Superior iron ores, 58
Laurel Furnace, 209n.8
Lee, Luke, 55, 200n.28
Lees, Jonathan, 68
Lesley, J. Peter, 62, 150
lime, Bluff Furnace, 137, 158
limonite: described, 12-13; in the Eastern Iron Belt of Tennessee, 44; in the Western Iron Belt of Tennessee, 44-45, 170; used in Citico Furnace, 169
Lincoln, Jesse, 48
Long, John P., 78, 199n.17
Long, Marcus B., 78
Lowe, Samuel B., 80
Lyon, Thomas C., 49, 53, 54, 199n.21

machine shops, 42
machinist, 75
magnetite: described, 13; in the Eastern Iron Belt of Tennessee, 44
Mann, R. D., 69
mantle or mantle plate, 24
masonry blast furnace, construction of, 21
mechanic, 27
Memphis and Charleston Railroad, 52
Menzi Muck Climbing Hoe, 93, 94 (illus.)
Middle Tennessee: early ironworks of, 46-47; iron production of, compared to East Tennessee, 56, 58
mold maker, 39
Montgomery, Benjamin R., 200n.38
Muscle Shoals, 50, 68

Narrows of Harpeth, 46, 168n.5
Nashville and Chattanooga Railroad, 50, 52, 62, 66, 72
National Register of Historic Places, 7
Negley, James S., 81

Overman, Frederick, 32, 36, 37, 150
Oxford Furnace, 162, 204n.76
oxidation, of iron, 134

Pactolus Ironworks, 46
Panic of 1857, 56
Paterson Forge, 47
photomicrography of iron samples, 148, 149, 150
pig iron: bar casting, 39; described, 11, 41
pig iron, Bluff Furnace. *See* Bluff Furnace, artifacts: pig iron
pig-iron production: 19th century statistics on, 18, 170, 171; U.S., 1850, 57; U. S., 1876, 170-72
Pioneer Mining and Manufacturing Company, 88
planters, 175-76
postmolds, 106
Potomac Furnace, coke-iron production at, 203n.77
Pratt, Milo, 84
Principio Company, 193n.11
profitability in southern manufacturing, 177
puddling furnace, 42

Raccoon Mountain, 66,
Raccoon Mountain Coal Company, 67, 84; *see also* Etna Mines
rail mill, 42
railroads: importance of iron with respect to, 52-53; serving Chattanooga, 50, 52; significance at Etna Mines, 66-67
Rankin, David, 66, 201-2n.56
reduction of iron, 35-37
refinery forge, 42
refractory index: Bluff Furnace estimates, 156, 158; defined, 158
research, archaeological, 95
retorts (for charcoaling), 16
reverberatory furnace, 42
risers, casting, 40, 143
Roane Iron Company, 170, 171
Robertson, James, 46
Rockwood, blast furnaces at, 170

Rockwood Formation, 150
Rockwood iron ores, 150, 152
Roddy coal banks (Rhea Co., Tenn.), 200-1n.40
Rogers, Spencer C., 55
rolling mill, 42
rolling mill, United State Military Railroad (Chattanooga), 83, 171
Ross, David, 46
Ross's Landing: Harry Fenn engraving of, 60; mentioned, 50, 52, 54, 61, 72
rotary blast fan. *See* blast machines
run-out arch, 23

salamander, 40; *see also* Bluff Furnace (archaeology): salamander
sampling, of artifacts,132
scaffolds, 40
Seely, Bruce, 163
self-fluxing iron ore, 13
Sevier, John, 46
Shelby Iron Works, 79, 86, 87-88
skip-hoist, 33
slag: chemistry and formation of, 36; defined, 7, 12, 38-39; formation of and removal in blast furnaces, 38; function of, 19; uses of, 39, 197n.34
slave insurrection panic of 1856, 76-77
slave labor: at Bluff Furnace, 76; fear of, 76-77, 176; gang organization of, as social control, 176-77; at Patterson Forge, 47; at southern industrial facilities, 175-77, 212n.44; at Tredegar Iron Works, 204n.90
smelting: by-products of, 38-39; description of, 11, 35; *see also* blast furnace
Society for Industrial Archeology, 2, 193n.1
Southern Iron Company, 171
Southern Manufacturing Company, 89
Southwestern Iron Company, 171
sow iron, 39
spectrographic analysis, 147-48
stack, blast furnace, 20
steam power, applied to blast furnaces, 27
stock, 12

stock house, 33
stock line, 37
stockpiles: defined, 12; importance of maintaining, 38
Strother, David Hunter, 64-65
Swan, William, 53-54, 199n.20

Tannehill Furnace (Alabama), 8, 88
tapping out, 38
Tennessee: antebellum ironworks, 45-47; charcoal-iron manufacture, 55-56; coal fields, 45; furnaces enumerated in Lesley (1859), 63-65; iron-producing regions, 44-45; iron-production statistics (1840-60), 56; mineral resources of, bibliographic, 197n.1; pig-iron production statistics (1876), 170-71
Tennessee, blast furnaces of: comparison between East and Middle Tennessee, 64-65; East Tennessee, enumerated (1859), 184-85; Middle Tennessee, enumerated (1859), 186-87
Tennessee Division of Archaeology, 8
Tennessee River: in East Tennessee physiography, 44; navigation problems on, 50; route of early European settlement, 45
throat. *See* tunnel head
tie rods, 21
timpstone, 21
titrimetric analysis, 147, 149
top gases: composition of, 36; recycling of heat from, 31-33, 196n.20
top men. *See* fillers
Treaty of New Echota, 47
Tredegar Iron Works, 76, 86
trompe (water blast), 26, 29
tunnel head, 20
tuyeres: cooling of, 30-31; described, 23; placement of, 29; water-cooled, 30, 32 (illus.)
tuyere arches, 23

Unaka Mountains, 44, 45
undercapitalization in southern industry, 175

United States Cast Iron Pipe and Foundry Company, 174
United States Military Railroad rolling mill (Chattanooga), 171
United States: iron production, 1840, 56; pig-iron production, 1850, 56, 57; pig-iron production, 1876, 172-73
Upland South iron-smelting region, 209n.2
urban sanitation, southern iron used for, 174
Ure, Andrew, 38, 195

Vulcan Iron Works, 80

Walnut Street Bridge, 7, 95, 182
water power, applied to blast furnaces, 26-27
Webster, Thomas, 68
Welcker, Albert G., 49
Wells, Moses, 67-68
Western and Atlantic Railroad, 50, 52, 72
Western Highland Rim: antebellum iron production on, 56-58; physiographic province, defined, 45; survey of iron sites on, 8
Western Iron Belt: defined, 44; postwar iron production in, 170; *see also* Tennessee, iron-producing regions
White, Dr. John R.: bibliography of, 207n.1; research of, 9, 131, 132, 157, 158
White's Creek, Roane County, Tenn. *See* Eagle Furnace
Whiteside, James Anderson: death of, 80; entrepreneurial focus of, 176; portrait of, 53 (illus.); real estate transactions of, 62; role in East Tennessee Iron Manufacturing Company, 54, 167; role in Etna Mines, 65-66
Wilder, John T., 81, 170, 171
Williams, William, 53, 166
workforce, of the blast furnace. *See* colliers, fillers, founder, guttermen, keeper, mechanic, mold makers, top men
wrought iron: general characteristics of, 20; manufacture of, 41, 42

Industry and Technology in Antebellum Tennessee: The Archaeology of Bluff Furnace
was designed by Kay Jursik and composed at The University of Tennessee
Press on the Apple Macintosh IIcx with Aldus *PageMaker*. Linotronic camera
pages were generated by AMPM, Inc. The book is set in Garamond No. 3 and
printed on 60-lb Glatfelter Natural, B-16. Manufactured in the United States
of America by Braun-Brumfield, Inc.

www.ingramcontent.com/pod-product-compliance
Lightning Source LLC
Chambersburg PA
CBHW031241290426
44109CB00012B/391